Extraterrestrial Life

Extraterrestrial Life

Tamara L. Roleff, *Book Editor*

David L. Bender, *Publisher*
Bruno Leone, *Executive Editor*
Bonnie Szumski, *Editorial Director*
Stuart B. Miller, *Managing Editor*
Brenda Stalcup, *Series Editor*

Contemporary Issues
Companion

Greenhaven Press, Inc., San Diego, CA

Every effort has been made to trace the owners of copyrighted material. The articles in this volume may have been edited for content, length, and/or reading level. The titles have been changed to enhance the editorial purpose. Those interested in locating the original source will find the complete citation on the first page of each article.

Library of Congress Cataloging-in-Publication Data

Extraterrestrial life / Tamara L. Roleff, book editor.
    p.   cm. — (Contemporary issues companion)
    Includes bibliographical references and index.
    ISBN 0-7377-0461-6 (pbk. : alk. paper). —
ISBN 0-7377-0462-4 (lib. bdg. : alk. paper)
    1. Life on other planets. I. Roleff, Tamara L., 1959– . II. Series.
QB54 .E945 2001
576.8'39—dc21   00-064045
                                                    CIP

©2001 by Greenhaven Press, Inc.
P.O. Box 289009, San Diego, CA 92198-9009

Printed in the U.S.A.

CONTENTS

Chapter 4: UFOs and Culture

FOREWORD

In the news, on the streets, and in neighborhoods, individuals are confronted with a variety of social problems. Such problems may affect people directly: A young woman may struggle with depression, suspect a friend of having bulimia, or watch a loved one battle cancer. And even the issues that do not directly affect her private life—such as religious cults, domestic violence, or legalized gambling—still impact the larger society in which she lives. Discovering and analyzing the complexities of issues that encompass communal and societal realms as well as the world of personal experience is a valuable educational goal in the modern world.

Effectively addressing social problems requires familiarity with a constantly changing stream of data. Becoming well informed about today's controversies is an intricate process that often involves reading myriad primary and secondary sources, analyzing political debates, weighing various experts' opinions—even listening to first-hand accounts of those directly affected by the issue. For students and general observers, this can be a daunting task because of the sheer volume of information available in books, periodicals, on the evening news, and on the Internet. Researching the consequences of legalized gambling, for example, might entail sifting through congressional testimony on gambling's societal effects, examining private studies on Indian gaming, perusing numerous websites devoted to Internet betting, and reading essays written by lottery winners as well as interviews with recovering compulsive gamblers. Obtaining valuable information can be time-consuming—since it often requires researchers to pore over numerous documents and commentaries before discovering a source relevant to their particular investigation.

Greenhaven's Contemporary Issues Companion series seeks to assist this process of research by providing readers with useful and pertinent information about today's complex issues. Each volume in this anthology series focuses on a topic of current interest, presenting informative and thought-provoking selections written from a wide variety of viewpoints. The readings selected by the editors include such diverse sources as personal accounts and case studies, pertinent factual and statistical articles, and relevant commentaries and overviews. This diversity of sources and views, found in every Contemporary Issues Companion, offers readers a broad perspective in one convenient volume.

In addition, each title in the Contemporary Issues Companion series is designed especially for young adults. The selections included in every volume are chosen for their accessibility and are expertly edited in consideration of both the reading and comprehension levels of the audience. The structure of the anthologies also enhances accessibility. An introductory essay places each issue in context and provides helpful facts such as historical background or current statistics and legislation that pertain to the topic. The chapters that follow organize the material and focus on specific aspects of the book's topic. Every essay is introduced by a brief summary of its main points and biographical information about the author. These summaries aid in comprehension and can also serve to direct readers to material of immediate interest and need. Finally, a comprehensive index allows readers to efficiently scan and locate content.

The Contemporary Issues Companion series is an ideal launching point for research on a particular topic. Each anthology in the series is composed of readings taken from an extensive gamut of resources, including periodicals, newspapers, books, government documents, the publications of private and public organizations, and Internet websites. In these volumes, readers will find factual support suitable for use in reports, debates, speeches, and research papers. The anthologies also facilitate further research, featuring a book and periodical bibliography and a list of organizations to contact for additional information.

A perfect resource for both students and the general reader, Greenhaven's Contemporary Issues Companion series is sure to be a valued source of current, readable information on social problems that interest young adults. It is the editors' hope that readers will find the Contemporary Issues Companion series useful as a starting point to formulate their own opinions about and answers to the complex issues of the present day.

Introduction

The possibility of the existence of extraterrestrial life has been debated for millennia. However, at the end of the nineteenth century, a wealthy amateur astronomer by the name of Percival Lowell thought he had settled the question once and for all. After an Italian astronomer named Giovanni Schiaparelli reported seeing *canali* (channels) on the surface of Mars, Lowell had an observatory built in 1894 near Flagstaff, Arizona. Using his powerful new telescope, he declared that Schiaparelli's *canali* were canals built by Martians to irrigate their dying planet. According to Lowell, the canals—averaging fifteen hundred miles in length—brought water from the polar ice caps to the planet's drought-stricken areas. Lowell found and named over four hundred canals on Mars and drew a detailed map showing their locations. Lowell also theorized that the surface of Mars had one-third the gravity of Earth; therefore, he asserted, the Martians were three times as large and twenty-seven times as strong as humans. In order to build the huge system of canals, he surmised, the Martians also must be much more intelligent than Earthlings.

Although stronger telescopes proved that Lowell's theory about canals on Mars was wrong, Lowell's observations changed the course of the discussion about extraterrestrial life. Instead of wondering *if* there was life on other planets, people now began wondering what those life forms might be like. One of the early descriptions of a human encounter with aliens appeared in 1898 in the book by H.G. Wells, *War of the Worlds*. When Orson Welles produced a radio play of *War of the Worlds* in 1936, the story of hostile, insectlike aliens attempting to conquer the Earth with their scorching death-ray terrified listeners who believed the invasion was actually taking place.

Accounts of encounters with aliens and unidentified flying objects were fairly rare until shortly after World War II. The modern era of flying saucers began in June 1947 when a commercial airline pilot named Kenneth Arnold, who was flying near Mount Rainier, Washington, saw "nine disks flying like a saucer skipped over the water." Just a few weeks later, one of the most famous events in UFO history occurred in Roswell, New Mexico, when a silvery object crashed in the desert. More than fifty years later, controversy still exists over whether the object was an alien spaceship or a secret military balloon.

Two months after the crash in Roswell, a Gallup poll found that nine out of ten Americans "knew about flying saucers," although few people reported that they had seen one or had come in contact with an alien species. Those who did say they had had a close encounter with extraterrestrials claimed that the aliens had come to Earth to collect soil samples and were extremely fearful of being seen by humans.

Descriptions of the aliens' appearance varied. Some witnesses reported that the aliens were indistinguishable from humans. Those who described the Roswell crash maintained that the ship's aliens were short (three to four feet tall) gray beings with oversized heads, large eyes, and only four fingers. This description became extremely common during the latter half of the twentieth century and spawned its own term to describe the alien race—the "grays."

Hollywood's version of aliens during the 1940s and 1950s featured two types of extraterrestrials—"benevolent space brothers" or "malevolent space monsters." According to Paul Meehan, author of *Saucer Movies: A UFOlogical History of the Movies,* the "benevolent space brothers" were portrayed as coming to Earth with messages of peace or to warn humans about the dangers of nuclear war. Early examples of this type of film include *The Day the Earth Stood Still* and *Stranger from Venus.* Terry Matheson, author of *Alien Abductions: Creating a Modern Phenomenon,* describes "malevolent space monsters" as "ugly, fearsome monsters" who were six to seven feet tall. Alien invasion movies such as *War of the Worlds, It Came from Outer Space,* and *Invaders from Mars* were popular with the public.

During the 1960s, most people who claimed they saw extraterrestrials said the aliens obeyed the laws of physics: For example, they used doors and walked up and down stairs. The sightings were usually in remote locations, Matheson notes, apparently due to the aliens' attempts to avoid detection by humans; contactees believed their sightings of aliens were random and unplanned. Contactees also said aliens displayed emotions, such as annoyance when asked questions by humans.

The first widely publicized account of an alien abduction appeared in 1966. In *The Interrupted Journey,* journalist John Fuller told the story of Betty and Barney Hill, who claimed they were abducted one night in September 1961 as they were driving home from a vacation in Canada. At a time when most sightings of flying saucers were, according to Matheson, "exciting visual mysteries," the Hills' story of abduction and experimentation by aliens was unusual and frightening. The Hills maintained that the aliens who abducted them resembled humans—"a red-headed Irishman," "a German Nazi" with "a black scarf around his neck," "a captain," and others, all of whom wore "dark jackets," according to Barney Hill.

Beginning in the 1970s, Matheson reports, extraterrestrials appeared to exhibit more malevolent behavior and attitudes toward those who allegedly had encounters with them. These abductees claimed that their alien captors seemed indifferent to any pain or fear their presence or experiments caused humans, he writes. In addition, aliens also began to possess supernatural powers during this time, such as being able to pass through walls, doors, and windows and traveling up toward the mother ship in beams of light. The aliens' ability to escape detection improved markedly; many abductees reported being taken while their spouse lay asleep next to them in bed or in the middle of bustling urban centers, such as New York City. According to Matheson, abductees started reporting that the aliens had implanted various devices inside their bodies. Abductees were now more likely to believe that they had been chosen deliberately and that the aliens continued to follow them for years. Some female abductees began to claim that they were being used as "breeders"; they described pregnancies that mysteriously disappeared and meetings with sickly looking alien children whom they felt were their own.

In the 1990s and up to the present day, Matheson relates, accounts of alien encounters have started to diverge into two distinct types: the malevolent aliens who conduct sometimes painful experiments on humans with little regard for their feelings, and benevolent aliens who leave their human contactees/abductees feeling spiritually energized. According to Matheson, some abductees assert that they have been chosen as prophets or messengers for their alien captors, "bringing warnings of impending catastrophe to humankind." They believe their extraterrestrial experience has made them more sensitive to the problems—environmental and otherwise—facing humans and the planet.

Hollywood's version of aliens during this period has also tended to depict the two different types of aliens, although the theme of malevolent aliens intent upon abductions, invasion, and destruction of the human race predominated, according to Meehan. Popular movies released during the 1990s featured abductions, such as *Intruders, Communion,* and *Fire in the Sky,* and invasions, such as *Independence Day, Starship Troopers,* and *Mars Attacks!* In *Contact,* however, aliens were depicted as spiritually benevolent beings eager to share their knowledge with humans.

As society has become more technologically advanced, so have the details contained in reports of alien encounters. The debate over whether extraterrestrial life even exists has evolved to include the questions of what such life might be like, whether aliens have visited Earth, and if so, what it is they want from humans. *Extraterrestrial Life: Contemporary Issues Companion* examines humans' ongoing obsession with the idea of aliens, presenting a selection of articles on the possibility of and search for life on other planets, close encounters with aliens, and how UFOs have influenced culture.

CHAPTER 1

IS THERE LIFE
BEYOND EARTH?

THE POSSIBILITY OF EXTRATERRESTRIAL LIFE

Joel Achenbach

In the following selection, Joel Achenbach, a writer for the *Washington Post* and the author of *Captured by Aliens: The Search for Life and Truth in a Very Large Universe*, provides an overview of the search for extraterrestrial life. He writes that although scientists have not yet discovered proof of life beyond Earth, they have found several tantalizing hints that it may exist somewhere in the universe. On the other hand, Achenbach notes that not all scientists are confident of finding extraterrestrial life, citing some discoveries and experiments that indicate that its existence is highly unlikely. However, Achenbach adds, as the technology to search for extraterrestrial life continues to improve, it is not inconceivable that eventually humans will discover some form of alien life.

Something astonishing has happened in the universe. There has arisen a thing called life—a flamboyant, rambunctious, gregarious form of matter, qualitatively different from rocks, gas, and dust, yet made of the same stuff, the same humdrum elements lying around everywhere.

Life has a way of being obvious—it literally scampers by, or growls, or curls up on the windowsill—and yet it's notoriously difficult to define in absolute terms. We say that life replicates. Life uses energy. Life adapts. Some forms of life have developed large central processing networks. In at least one instance, life has become profoundly self-aware.

And that kind of life has a big question: What else is alive out there?

There may be no scientific mystery so tantalizing at the brink of the new millennium and yet so resistant to an answer. Extraterrestrial life represents an enormous gap in our knowledge of nature. With instruments such as the Hubble Space Telescope, scientists have discovered a bewildering amount of cosmic turf, and yet they still know of only a single inhabited world.

We all have our suppositions, our scenarios. The late astronomer Carl Sagan estimated that there are a million technological civilizations in our galaxy alone. His more conservative colleague Frank Drake offers the number 10,000. John Oro, a pioneering comet researcher,

Excerpted from Joel Achenbach, "Life Beyond Earth," *National Geographic,* January 2000. Reprinted by permission of the National Geographic Society.

calculates that the Milky Way is sprinkled with a hundred civilizations. And finally there are skeptics like Ben Zuckerman, an astronomer at UCLA [the University of California at Los Angeles], who thinks we may well be alone in this galaxy if not in the universe.

All the estimates are highly speculative. The fact is that there is no conclusive evidence of any life beyond Earth. Absence of evidence is not evidence of absence, as various pundits have wisely noted. But still we don't have any solid knowledge about a single alien microbe, a solitary spore, much less the hubcap from a passing alien starship.

Our ideas about extraterrestrial life are what Sagan called "plausibility arguments," usually shot through with unknowns, hunches, ideologies, and random ought-to-bes. Even if we convince ourselves that there must be life out there, we confront a second problem, which is that we don't know anything about that life. We don't know how truly alien it is. We don't know if it's built on a foundation of carbon atoms. We don't know if it requires a liquid-water medium, if it swims or flies or burrows.

Despite the enveloping nebula of uncertainties, extraterrestrial life has become an increasingly exciting area of scientific inquiry. The field is called exobiology or astrobiology or bioastronomy—every few years it seems as though the name has been changed to protect the ignorant.

Whatever it's called, this is a science infused with optimism. We now know that the universe may be aswarm with planets. Since 1995 astronomers have detected at least 22 planets orbiting other stars. NASA hopes to build a telescope called the Terrestrial Planet Finder to search for Earth-like planets, examining them for the atmospheric signatures of a living world. In the past decade organisms have been found thriving on our own planet in bizarre, hostile environments. If microbes can live in the pores of rock deep beneath the earth or at the rim of a scalding Yellowstone spring, then they might find a place like Mars not so shabby.

Mars is in the midst of a full-scale invasion from Earth, from polar landers to global surveyors to rovers looking for fossils. A canister of Mars rocks will be rocketed back to Earth in the year 2008, parachuting into the Utah desert for scrutiny by scientists in a carefully sealed lab. In the coming years probes will also go around and, at some point, into Jupiter's moon Europa. That icy world shows numerous signs of having a subsurface ocean—and could conceivably harbor a dark, cold biosphere.

The quest for an alien microbe is supplemented by a continuing effort to find something large, intelligent, and communicative. SETI—the Search for Extraterrestrial Intelligence—has not yielded a confirmed signal from an alien civilization in 40 years of experiments, but the signal-processing technology grows more sophisticated each year. The optimists figure it's only a matter of time before we tune in the right channel.

No one knows when—or if—one of these investigations might make a breakthrough. There's a fair bit of boosterism surrounding the entire field, but I'd bet the breakthrough is many years, if not decades, away. The simple truth: Extraterrestrial life, by definition, is not conveniently located.

But there are other truths that sustain the search for alien organisms. One is that, roughly speaking, the universe looks habitable. Another is that life radiates information about itself—that, if nothing else, it usually leaves a residue, an imprint, an echo. If the universe contains an abundance of life, that life is not likely to remain forever in the realm of the unknown.

Contact with an alien civilization would be an epochal and culturally challenging event, but exobiologists would settle gladly for the discovery of a tiny fossil, a mere remnant of extraterrestrial biochemistry. One example. One data point to add to the one we have—Earth life. That's what we need to begin the long process of putting human existence in its true cosmic context. . . .

The most tantalizing possibility is that the universe hums with life and that in the coming centuries we will find it. An exobiologist's abiding optimism is fired by the knowledge that living things are primarily constructed of hydrogen, nitrogen, carbon, and oxygen—the four most common chemically active elements in the universe. And life is inextricably interwoven with nonlife; not even the sharpest razor can perfectly slice them apart.

We also know that a functioning ecosystem does not require sunlight or photosynthesis. In the early 1990s researchers found that the basaltic rock deep beneath Washington State contains an abundance of microbes totally cut off from the photosynthetic world. Even more complex life can adapt to hostile places. When scientists in the deep-sea submersible *Alvin* went tooling around the mid-ocean ridges, they found hot vents covered with shrimp and mouthless tube worms.

What remains unknown is whether life can survive over time in narrow ecological niches on largely barren worlds. Could life survive in aquifers far below the harsh surface of Mars? What could endure the cold, dark environment of Europa's hypothesized ocean? Can an alien world have just a little bit of life, or are biospheres an all-or-nothing proposition? . . .

Our contemporary culture did not invent this idea of life beyond Earth. The alien is a Hollywood stock character but not a Hollywood creation. More than 2,000 years ago the Greek philosopher Metrodorus of Chios wrote, "It is unnatural in a large field to have only one shaft of wheat, and in the infinite Universe only one living world." Four centuries ago Giordano Bruno was burned at the stake in part because he believed that there were inhabited worlds throughout the cosmos. Astronomers like Christian Huygens supplemented their purely scientific work with treatises on the characteristics of life beyond Earth. Huygens felt, for example, that aliens would probably have hands, like humans.

Missing from the debate, typically, was the one ingredient of a truly persuasive argument: Evidence. That seemed to change with the apparent discovery of the Martian canals. In 1877 Giovanni Schiaparelli, an Italian astronomer, found what he called canali, or channels, on the surface of the planet. The American astronomer Percival Lowell and a few colleagues took the idea from there.

In the final years of the 19th century, Lowell, using a new telescope he built near Flagstaff, Arizona, revealed the discovery of hundreds of canals and argued that these were the artificial creations of an intelligent Martian civilization. In fact, he wrote, the Martians would certainly have to be superior to us. He reasoned that their globe-spanning engineering projects were far beyond our own capabilities and that the ability of a race of creatures to live in harmony over the whole of a planet showed them to be of a more advanced character than our own squabbling selves. H.G. Wells tweaked the idea just a bit in his novel *The War of the Worlds*, in which the Martians come to Earth with deadly heat rays and dreams of conquest.

The Martians, alas, were doomed, except as cultural artifacts. When astronomers looked at Mars with more powerful telescopes, there were no canals anywhere. Lowell's canals were created in his mind's eye—a classic example of the saying "Believing is seeing." But there remained, into the 1960s, a fascination with waves of seasonal darkening on the surface. Could this be vegetation? The Martian prairies and forests were conclusively eradicated in 1965, when the Mariner 4 probe took 22 pictures of the surface. Mars was a cratered wasteland, reminiscent of the moon.

When the Viking landers descended to the Martian surface in 1976, they found no compelling sign of life and indeed discovered that the surface contains no trace of organic molecules. Though the mission was a fantastic triumph of science and technology, the absence of detectable life on Mars put exobiology into a two-decade funk.

The mood changed in the 1990s. Biologists were detecting organisms in such exotic environments on Earth that they were inspired to look anew at the rest of the solar system as potentially habitable. They also discovered signs that life appeared early in the Earth's history. Intriguingly, at about the time life arose on Earth, Mars was a much more hospitable planet than it is today. Images of the Martian surface indicate that the planet once had flowing rivers and perhaps an ocean. Life could even have started on Mars and spread to Earth aboard a meteorite.

Which brings up the most famous Martian meteorite: ALH84001. In 1996 a team of three NASA scientists based in Houston announced that this potato-size rock, found in Antarctica, contained what appeared to be Martian fossils.

The discovery was proclaimed at an unforgettable NASA press conference in Washington, D.C., on August 7, 1996. Everyone realized the historical glory of being right about these purported microfossils—and the reciprocal tarnish of being wrong. Dan Goldin, NASA Administrator, cautioned that the results were not definitive, but he said, "We may see the first evidence that life might have existed beyond the confines of this planet, the third rock from the sun."

The NASA team made a dramatic presentation, complete with graphics and the first, startling images of the microfossils, one of which looked like a worm (others a bit like Cheetos). But then came a dissenter, UCLA's J. William Schopf, who said that on a scale of one to ten of increasing probability of biological origin, he could only grant the alleged Martian fossils a two. So began, that day, an enduringly divisive scientific debate.

The NASA scientists had to admit that their four primary lines of evidence could each be explained nonbiologically. They had found, for example, PAHs, polycyclic aromatic hydrocarbons, which sometimes are associated with living things but which also can be found in car exhaust. They found grains of magnetite, which might have been produced inside microbes or might not have. In a sense the research raised the question of whether a series of possibilities add up to a probability. At the least it runs headlong into a Sagan dictum, which is that extraordinary claims require extraordinary evidence.

The NASA team saw its conclusions vigorously attacked. One damaging study showed that some of the microbe-like structures were merely flakes of the rock rendered more biological in appearance by the coating process used in the preparation of slides. Researchers also found contaminants from Earth inside the meteorite.

The team fought these challenges point by point, but after three years critics felt they'd pretty much killed off the Mars rock. Luann Becker, a geochemist at the University of Hawaii, told me, "I think we're beating a dead horse."

But Everett Gibson, part of the NASA meteorite team, sees this as a typical scientific resistance to a revolutionary idea. "Science," he said, "doesn't accept radical ideas quickly."

There was a time when scientists didn't believe that meteorites could possibly fall from the sky. There was a time when plate tectonics—the movement and collision and subduction of vast slabs of the Earth's crust—was deemed a very strange idea. Are the Mars rock fossils in the same category? Or are they more like those canals?

If life sprang up through natural processes on the Earth, then the same thing could presumably happen on other worlds. And yet when we look at outer space, we do not see an environment teeming with life. We see planets and moons where no life as we know it could possibly survive. In fact we see all sorts of wildly different planets and moons—hot places, murky places, ice worlds, gas worlds—

and it seems that there are far more ways to be a dead world than a live one.

Within our solar system the Earth may be in a fairly narrow habitable zone, not too hot and not too cold, just the right distance from the sun that water can splash around on the surface in a liquid state. And there may be many other things that make life on Earth possible. The tectonic activity recycles the planet's carbon. Mars has no such mechanism, and this seemingly minor deficiency may be the reason Mars lost most of its atmosphere.

The search for extraterrestrial life is in some ways a search for constraints, for the things that limit the emergence of life or the evolution of complex organisms. For calculating the number of technological, communicative civilizations, the most popular theoretical tool is the Drake equation.

In 1960 an American astronomer named Frank Drake became the first person to conduct a sensitive radio search for signals from extraterrestrial civilizations. He aimed an 85-foot radio telescope at two nearby stars and, after one false alarm, found no intentional signals. The next year, preparing for a meeting of visionary thinkers (including the young Sagan), he made an outline for how to discuss the probability of detecting intelligent life, starting with the rate of star formation and the typical number of planets and working through to the longevity of civilizations. "I thought it was just a gimmick. It's amazing to me now that it's in the astronomy textbooks," he told me.

Going through the factors from left to right—$N = R_* f_p n_e f_l f_i f_c L$—you don't get very far before you hit some serious unknowns. Jill Tarter, who has dedicated her career to SETI, says, "The Drake equation is a wonderful way to organize our ignorance."

The only factor well understood, R_*, tells us the number of stars. Suffice it to say that there are lots of them, more than a hundred billion in our galaxy alone, maybe as many as 400 billion (and that doesn't count, of course, the billions of other galaxies). The second factor, f_p, the fraction of stars with planets, is rapidly coming into focus. There are still uncertainties, since the detection equipment can find only extremely massive planets. These behemoths aren't like the Earth. Many of the extrasolar planets discovered so far may have migrated toward the parent star over time, destroying any rocky, Earth-like planets along the way.

Eventually the Terrestrial Planet Finder (TPF) could help solve the next factor in the equation, n_e—the number of planets with habitable environments—and may even be able to glean some evidence of the following factor, f_l—the fraction on which life has originated. TPF, still many years from construction, would capture the feeble reflected light from a distant rocky planet, while nulling the far more brilliant light of the parent star. This remnant of light might amount to only a single pixel of data. The light could then be examined for the spectral signatures of, for example, oxygen, methane, ozone, or some indicator of a planet with biological processes. Thrilling as such a discovery would be, it's easy to imagine how it would echo the situation with the Mars rock. There would likely be no "proof" of life, merely an interpretation subject to much second-guessing.

Even on Earth the origin of life is a stubbornly enduring mystery. "How can a collection of chemicals form themselves into a living thing without any interference from outside?" asks Paul Davies, a physicist and writer. "On the face of it, life is an exceedingly unlikely event," he argues. "There is no known principle of matter that says it has to organize itself into life. I'm very happy to believe in my head that we live in a biofriendly universe, because in my heart I find that very congenial. But we have not yet discovered the Life Principle."

No one is even sure that life requires liquid water, though that seems a reasonable bet and is surely the case on Earth. Liquid water may be fairly scarce in the universe—Europa may help solve that issue—but another presumed ingredient of life, organic molecules, those made up primarily of carbon, are commonplace. That's why Jeffrey Bada, a pretty hard-nosed researcher, thinks the universe is full of living things. "I don't see any way to avoid that," he said, sounding almost apologetic.

So let's assume that life can spring up in many places. Now comes f_i, another giant unknown in the Drake equation: How often does life evolve to a condition of intelligence?

There are those, like Ernst Mayr, one of the great biologists of the 20th century, who argue that high intelligence has occurred only once on Earth, among something like a billion species. Hence it is a billion-to-one long shot. But Paul Horowitz, a Harvard physicist, argues that the same data can be looked at the opposite way: That on the only planet we know of that has life, intelligence appeared. That's a one-for-one proposition.

I've never met anyone who thinks that if you rewound the tape of terrestrial evolution (to use Stephen Jay Gould's metaphor) and played it again, you'd wind up with a genetically identical human being the second time around. But there are those who say that an intelligent being is more likely under certain initial conditions. The paleobiologist Andy Knoll argues that intelligence is rooted in the emergence of structures that allow simple animals to sense their environment and seek food. "If we get to creepy crawlies that look for food, then at some point intelligent life may emerge," he says.

There are those who argue passionately that alien life would be nothing like us—in Fred Hoyle's novel *The Black Cloud* the alien is a gaseous cloud that decides to feed on our sun—and there are others who say the biology of the Earth is probably a pretty good example of what's out there.

Finding life somewhere else, even a single alien amoeba, might clarify the extent to which life evolves along parallel tracks—and whether it typically arrives at certain useful structures, such as eyeballs, wings, and large brains. Human beings have, by far, the biggest brains on Earth in ratio to body size. Did we get these things in our skulls through a random, improbable evolutionary quirk?

Lori Marino, a psychobiologist at Emory University, points out that dolphins appear to have undergone a dramatic increase in brain size in the past 35 million years, which may have a parallel in the quadrupling of brain size among hominids in the past few million years. By her reckoning, huge leaps in intelligence may be found among creatures on worlds everywhere else in the universe.

But it's also true that the data are scarce, and this is still a territory for, among others, philosophers and theologians. What does it mean to be "intelligent"? When we "think" or "feel" or "love," what is it that we are doing? When we ask if we are "alone," we really want to know if there are others out there in the universe who are, in key aspects, very much like ourselves. We seek the communicators—Drake's f_c, creatures who have the technology to send signals—storytellers, ideally. . . .

Freeman Dyson, a physicist, has argued that humans may engineer new forms of life that will be adapted to living in the vacuum of space or on the surface of frozen moons and comets and asteroids. In Dyson's universe, life is mobile, and planets are gravitational traps inhibiting free movement.

"Perhaps our destiny is to be the midwives, to help the living universe to be born," he said recently. "Once life escapes from this little planet, there'll be no stopping it."

But life must first survive this planet. The longevity of civilizations is the final factor in the Drake equation, the haunting letter L. Humans in their modern anatomy have been around only 125,000 years or so. It is not clear yet that a brain like ours is necessarily a long-term advantage. We make mistakes. We build bombs. We ravage our world, poison its water, foul its air. Our first order of business, as a species, is to make L as long an interval as possible.

I would hope that anyone who investigates this issue will come away with a renewed appreciation of what and who we are. In a universe of empty space and stellar furnaces and ice worlds, it is good to be alive. And we should remember that even if we find intelligent life beyond Earth, it may not be what we expect or even what we were searching for.

The alien may not speak to that part of our consciousness that we deem most important—our spirit, if you will. It may have little to teach us. The great moment of contact may simply remind us that what we most want is to find a better version of ourselves—a creature we will probably have to make, from our own raw elements, here on Earth.

WE ARE ALONE

Guillermo Gonzalez

There is a reason why no evidence of extraterrestrial intelligence has been found, Guillermo Gonzalez asserts in the following article. The complex conditions needed to create and sustain life could only be present on an extremely limited number of planets, he contends, and therefore Earth's humans are most likely the sole form of intelligent life in the universe. According to the author, since the possibility of extraterrestrial life is highly unlikely, humans should not concentrate on asking "Are we alone?" but instead "Why are we here?" Gonzalez is a research astronomer at the University of Washington.

Are we alone? It's the question we astronomers are most frequently asked. Fascinated as always by the notion of extraterrestrial intelligence, people are flocking to see "Men in Black" and "Contact" and watching with wonder Pathfinder's images from Mars.

My answer to the question almost always catches people off guard: Very likely yes, we are alone. When one looks at the astronomical data with an open mind, it becomes quite obvious why we have not found any evidence of extraterrestrial intelligence.

The subject was not seriously addressed by astronomers until the 1960s. In 1966 Carl Sagan and Iosef Shklovski estimated that the Milky Way had a million habitable planets. But they based this on just two astronomical constraints: type of parent star and distance from that star. As knowledge of the universe has increased, astronomers have quietly been adding to the list of astronomical constraints. It is now clear that the early estimates were wildly optimistic.

The first step in calculating the probability of extraterrestrial intelligence is determining the biological requirements for life. Given that the laws of chemistry and physics are universal, we can be fairly certain about inferring the essentials from our observations on Earth. The most basic ingredients are liquid water, a long-term, stable energy source and a number of key chemical elements (bacteria require hydrogen, carbon, nitrogen, oxygen, sodium, magnesium, phosphorous, sulfur, chlorine, potassium, calcium, manganese, iron, cobalt, copper, zinc and molybdenum; humans require some 10 additional elements).

Of course, just having these basic ingredients does not guarantee life; one must follow the directions carefully when baking a cake.

To determine the probability of intelligent life existing elsewhere in the Milky Way, we must consider the necessary conditions for both the origin of life and the maintenance of life. Examples of the first category: a source of heavy elements, required to build terrestrial planets and organisms (this rules out most old stars, which formed from hydrogen and helium); a source of radioactive elements to keep the core of a planet hot (this rules out recently formed stars); and a source of liquid water (which may come from comets). These requirements limit the possible locations in the Milky Way where life might originate. Heavy elements are more plentiful in the galaxy's inner regions, and terrestrial planets could not have formed early in the Milky Way's history, because the concentration of heavy elements was too low. There exists only a small "window of opportunity"—both in space and time—for life to form in the Milky Way.

The conditions needed to sustain life are different from—and in some cases seem to contradict—those needed to form it. Among them: the presence of a large moon to stabilize a planet's axial tilt and slow down its rotation rate; the absence of nearby novae and supernovae; a circular orbit in the plane of the Milky Way; and the presence of a "gas giant"—a Jupiter-like planet—to regulate the influx of comets. I could list well over two dozen other astronomical constraints.

Recent discoveries in astronomy greatly weaken the case for extraterrestrial life by demonstrating how dangerous the universe is. Gamma-ray bursts turn out to give off much more radiation than had previously been believed. Using the Hubble Space Telescope, astronomers have observed that most nearby large galaxies harbor massive black holes in their centers, which also give off prodigious quantities of dangerous radiation. And the link between the meteor crater near Chicxulub, Mexico, and the demise of the dinosaurs shows how vulnerable existing life is to planetary catastrophe.

But what about recent discoveries in our own solar system—the evidence of life reported on a Martian rock last year, and indirect evidence of an ocean under the frozen surface of Jupiter's moon Europa? Additional research on the Mars rock has greatly weakened the case for Martian life. And the possible existence of liquid water below Europa's surface is far from a guarantee of life. The lesson we should learn from the spectacular pictures coming back from the latest Mars mission is not that life might be present, but that even for the most Earth-like planet in the solar system, the conditions appear too harsh even for simple life.

We should not be asking: "Are we alone?" We should be asking instead: "Why are we here?"

THE POSSIBILITY OF COMPLEX EXTRATERRESTRIAL LIFE IS REMOTE

Peter D. Ward and Donald Brownlee

The odds that complex life-forms exist elsewhere in the universe are exceedingly remote, contend Peter D. Ward and Donald Brownlee in the following selection. The authors believe that the factors that contributed to the evolution and survival of intelligent life on Earth—such as the size and intensity of the sun, the particular formation of Earth and the solar system, the presence of the Moon and Jupiter—are simply too random and rare to have occurred on other planets. Scientific predictions that the galaxy holds millions of planets with intelligent life are too optimistic, they conclude. Ward is a paleontologist who researches mass extinctions. Brownlee is an astronomer. They are the authors of *Rare Earth: Why Complex Life Is Uncommon in the Universe,* from which this essay is excerpted.

How rare is Earth? . . . A long grocery list of ingredients [is] seemingly necessary to make a planet teeming with complex life. It involves material, time, and chance events. In this essay we will try to assess these various factors and their relative importance; all can be thought of as probabilities. In some cases we understand these probabilities, but in others almost no research has been done, and our questions are the simple questions of children—questions as yet with no answers. Some of these questions can thus be tackled only with our imagination. Others will be answered by [information from] space voyages and instrumented investigations.

The Formation of a Suitable Star

Let us begin by imagining we have the power to observe 100 solar nebulae coalesce into stars and the planets that will encircle them. How many of these events will yield an Earth-like planet with animal life?

The first step in preparing the way for a habitable environment is the formation of a suitable star: one that will burn long enough to let evolution work its wonders, one that does not pulse or rapidly change its energy output, one without too much ultraviolet radiation, and most important, perhaps, one that is large enough. Of the 100 applicants, perhaps only two to five will yield a star as large as our sun. The vast majority of stars in the Universe are smaller than our sun, and although smaller stars could have planets with life, most would be so dim that Earth-like planets would have to orbit very close to their star to receive energy sufficient to melt water. But being close enough to get adequate energy from a small star leads to another problem: tidal lock, the condition where the same side of the planet always faces the sun. A tidally locked planet is probably unsuitable for animal life.

What if we increased the number to 1000 planetary systems, so that we might expect 20 stars of our sun's size or greater to be born? Even these numbers are too small to yield a high probability that we will find a truly Earth-like planet. Perhaps a better way to envision the various odds is to re-create the scenario that led to the formation of our solar system and then run through the process once again in a thought experiment. Stephen Jay Gould used this type of mental reconstruction in his interpretation of the Cambrian Explosion. In his 1989 book *Wonderful Life,* Gould described the exercise as follows:

> I call this experiment "replaying the tape." You press the rewind button and, making sure you thoroughly erase everything that actually happened, go back to any time and place in the past—say, to the seas of the Burgess Shale. Then let the tape run again and see if the repetition looks at all like the original.

A Thought Experiment

In our case, we will replay the tape of our planet's formation. We begin with a planetary nebula of exactly the mass and elemental composition that created our solar system. According to most theorists, this might create a star identical to our sun—but then it might not. For instance, the spin rate of the new star might be different from that of our sun, with unknown consequences. Well, then, 1000 such solar nebulae, *perhaps* 1000 clones of dear old Sol. Not so, however, with the planets coalescing out of this mix. If we rerun this particular tape, we will in all probability not get a repeat of our solar system with its nine planets, its one failed planet (now the asteroid belt), a Jupiter and three other gas giants orbiting outside four terrestrial planets; and a halo of comets surrounding the entire mix. Now we enter the realm of multiple contingencies. Of the 1000 newly formed planetary systems, none is likely to be identical to our solar system today—just as no two people are identical. In a coalescing planetary system many processes, including planetary formation, may be chaotic.

Planets form in what are known as "feeding zones," regions where various elements come together and eventually coalesce into planetesimals, which finally aggregate into a planet. Recent work by planetary scientists shows that the spacing of planets will probably be fairly regular. There might be as few as six planets or as many as ten or even more. James Kasting of Penn State University believes that planetary spacing is not accidental—that the positions of planets are highly regulated, and that if the solar system were to re-form many times, we would get the same number of planets each time. Yet the observational evidence to date does not back up the theory. The extrasolar planets that have been discovered exhibit an enormous diversity of spacing and orbits; their positions are not nearly so orderly as the theory suggests they should be. Ross Taylor, an astronomer who received the prestigious Leonard Award in 1998, disputes Kasting's views. "Clearly," he maintains, "the conditions that existed to make our system of planets are not easily reproduced. Although the processes of forming planets around stars are probably broadly similar, the devil is in the details."

No one knows whether a planet the size of Jupiter would always form or whether there would be a couple of planets like Mars instead. A planet would probably form in about the position of Earth, but it could be larger or smaller, somewhat closer to the sun or farther away. Would the material (physical) quantities be essentially the same? Would plate tectonics develop? Would there be the same amount of water—and would that water end up on the surface of the planet, rather than locked up in its mantle or lost to space? Would there be few threats to life from Earth orbit–crossing asteroids? What is the chance that our Moon would form again, if, as we believe, it is important in making Earth a stable place conducive to animal diversification?

Even if all of these events occurred more or less the way they have, would life form again? And given life, would animal life appear once more? Can there be animal life without the utterly chance events that occurred in Earth's history? . . .

Let us reorganize (and rephrase) this set of questions in the following way. We might ask: How many of all planets in the Universe are terrestrial planets (as opposed to the gas giants such as Jupiter, for instance)? What is the percentage for all planets of the Universe? (In our solar system there are five, but if we add the larger moons, that number more than triples.) Of the terrestrial planets in the Universe, how many have enough water to form an ocean (either as water or as ice)? Of those planets with oceans, how many have any land? Of those with land, how many have continents (rather than, say, scattered islands)? Yet these questions are only for the infinitesimally small slice of time we call the present. All of these conditions are subject to change. . . .

The Importance of Our Large Moon

Although many scientists have been doggedly pursuing the various attributes necessary for a habitable planet—Michael Hart, George

Weatherill, Chris McKay, Norman Sleep, Kevin Zahnlee, David Schwartzman, Christopher Chyba, Carl Sagan, and David Des Marais come to mind—one name stands out in the scientific literature: James Kasting of Penn State University.

Kasting notes that whether habitable planets exist around other stars "depends on whether other planets exist, where they form, how big they are, and how they are spaced." Kasting stresses, as we do, the importance of plate tectonics in creating and maintaining habitable planets, and he suggests that the presence of plate tectonics on any planet can be attributed to the planet's composition and position in its solar system. But one of Kasting's most intriguing comments is related to our Moon. Kasting notes that the obliquity (the angle of the axis of spin of a planet) of three of the four "terrestrial" planets of our solar system—Mercury, Venus, and Mars—has varied chaotically.

> Earth is the exception, but only because it has a large moon. . . . If calculations about the obliquity changes in the absence of the moon are correct, Earth's obliquity would vary chaotically from 0 to 85 degrees on a time scale of tens of millions of years were it not for the presence of the Moon. . . . Earth's climatic stability is dependent to a large extent on the existence of the Moon. The Moon is now generally believed to have formed as a consequence of a glancing collision with a Mars-sized body during the later stages of the Earth's formation. If such moon-forming collisions are rare . . . habitable planets might be equally rare.

We have accumulated a laundry list of potentially low-probability events or conditions necessary for animal life: not only Earth's position in the "habitable zone" of its solar system (and of its galaxy), but many others as well, including a large moon, plate tectonics, Jupiter in the wings, a magnetic field, and the many events that led up to the evolution of the first animal. Let us explore what these conditions might mean for life beyond Earth.

The Odds of Animal Life Elsewhere

In the 1950s, astronomer Frank Drake developed a thought-provoking equation to predict how many civilizations might exist in our galaxy. The point of the exercise was to estimate the likelihood of our detecting radio signals sent from other technologically advanced civilizations. This was the beginning of sporadic attempts by Earthlings to detect intelligent life on other planets. Now called the Drake Equation in its creator's honor, it has had enormous influence in a (perhaps necessarily) qualitative field. The Drake Equation is simply a string of factors that, when multiplied together, give an estimate of the number of intelligent civilizations, N, in the Milky Way galaxy.

As originally postulated, the Drake Equation is:

$$N^* \times fs \times fp \times ne \times fi \times fc \times fl = N$$

where:

N^* = stars in the Milky Way galaxy
fs = fraction of sun-like stars
fp = fraction of stars with planets
ne = planets in a star's habitable zone
fi = fraction of habitable planets where life does arise
fc = fraction of planets inhabited by intelligent beings
fl = percentage of a lifetime of a planet that is marked by the presence of a communicative civilization

Our ability to assign probable values to these terms varies enormously. When Drake first published his famous equation, there were great uncertainties in most of the factors. There did (and does) exist a good estimate for the number of stars in our galaxy (between 200 and 300 million). The number of star systems with planets, however, was very poorly known in Drake's time. Although many astronomers believed that planets were common, there was no theory that *proved* star formation should include the creation of planets, and many believed that the formation of planetary systems was exceedingly rare. During the 1970s and later, however, it was assumed that planets were common; in fact, Carl Sagan estimated that an average of ten planets would be found around *each star*. Even though no extrasolar planets were found until the 1990s, their discovery seemed to vindicate those who believed planets were common. But is it so? A new look at this problem suggests that planets may indeed be quite rare—and thus the presence of animal life rarer still.

Are Stars with Planets Anomalous?

We now know that planetary formation outside our own system does indeed occur. The recent and spectacular discovery of extrasolar planets, one of the great triumphs of astronomical research in the 1990s, has proved what has long been assumed: that other stars have planets. But at what frequency? It may be that a substantial fraction of stars have planetary systems. To date, however, astronomers have succeeded in detecting only giant, "Jupiter-like" planets; available techniques cannot yet identify the smaller, rocky, terrestrial worlds. Now that numerous stars have been examined, it appears that *only about 5% to 6% of examined stars have detectable planets.* Because only large gas-giant planets can be detected, this figure really shows that Jupiter clones close to stars or in elliptical orbit are rare. But perhaps it indicates that planets as a whole are rare as well.

The evidence that planets may be rare comes not so much from the direct-observation approach of the planet finders but from spectroscopic studies of stars that appear similar to our own sun. The studies of those stars around which planets have been discovered have yielded an intriguing finding: They, like our sun, are rich in metals. According to astronomers conducting these studies, there seems to be a causal link between high metal content in a star and the presence of planets. Our own star is metal-rich. In a study of 174 stars, astronomer G. Gonzalez discovered that the sun was among the highest in metal content. It appears that we orbit a rare sun.

Other new studies also require us to question the belief that planetary systems such as our own are common. At a large meeting of astronomers held in Texas in early 1999, it was announced that 17 nearby stars had been observed to be orbited by planets the size of Jupiter. Astronomers at the meeting were also puzzled by an emerging pattern: None of the extrasolar planetary systems resembles the sun's family of planets. Geoff Marcy, the world's leading planet finder, noted that "for the first time, we have enough extrasolar planets out there to do some comparative study. We are realizing that most of the Jupiter-like objects far from their stars tool around in elliptical orbits, not circular orbits, which are the rule in our solar system." All of the Jupiter-sized objects either were found in orbits much closer to their sun than Jupiter is to our sun, or, if they occurred at a greater distance from their sun, had highly elliptical orbits (observed in 9 of the 17 so far detected). In such planetary systems, the possibility of Earth-like planets existing in stable orbits is low. A Jupiter close to its sun will have destroyed the inner rocky planets. A Jupiter with an elliptical or decaying orbit will have disrupted planetary orbits sunward, causing smaller planets either to spiral into their sun or to be ejected into the cold grave of interstellar space.

It is still impossible to observe smaller, rocky planets orbiting other stars. Perhaps such planets—which we believe are necessary for animal life—are quite common. But perhaps this is a moot point. We have hypothesized that animal life cannot long exist on a planet unless there is a giant, Jupiter-like planet within the same planetary system—and orbiting outside the rocky planets—to protect against comet impacts. It may be that Jupiters like our own, in regular orbits, are rare as well. To date, all tend to be in orbital positions that would be lethal, rather than beneficial, to any smaller, rocky planets.

Planet Frequency and the Drake Equation

All predictions concerning the frequency of life in the Universe inherently assume that planets are common. But what if the conclusions suggested by emerging studies—that Earth-like planets are rare, and planets with metal rarer still—are true?

This finding has enormous significance for the final answer to the Drake Equation. Any factor in the equation that is close to zero yields a near-zero final answer, because all the factors are multiplied together.

Carl Sagan, in 1974, estimated that the average number of planets around each star is ten. Donald Goldsmith and Tobias Owen, in their 1992 *The Search for Life in the Universe,* also estimated ten planets per star. But the new findings suggest greater caution. Perhaps planetary formation is much less common than these authors have speculated.

To estimate the frequency of intelligent life, the Drake Equation hinges on the abundance of Earth-like planets around sun-like stars. The most common stars in the galaxy are M stars, fainter than the sun and nearly 100 times more numerous than solar-mass stars. These stars can generally be ruled out because their "habitable zones," where surface temperatures could be conducive to life, are uninhabitable for other reasons. To be appropriately warmed by these fainter stars, planets must be so close to the star that tidal effects from the star force them into synchronous rotation. One side of the planet always faces the star, and on the permanently dark side, the ground reaches such low temperatures that the atmosphere freezes out. Stars much more massive than the sun have stable lifetimes of only a few billion years, which might be too short for the development of advanced life and evolution of an ideal atmosphere. Each planetary system around a 1-solar-mass star will have *space* for at least one terrestrial planet in its habitable zone. But will there actually *be* an Earth-sized planet orbiting its star in that space? When we take into account factors such as the abundance of planets and the location and lifetime of the habitable zone, the Drake Equation suggests that only between 1% and 0.001% percent of all stars might have planets with habitats similar to those on Earth. But many now believe that even these small numbers are overestimated. On a universal viewpoint, the existence of a galactic habitable zone vastly reduces them.

Such percentages seem very small, but considering the vastness of the Universe, applying them to the immense numbers of stars within it can still result in very large estimates. Carl Sagan and others have mulled these various figures over and over. *They ultimately arrived at an estimate of one million civilizations of creatures capable of interstellar communication existing in the Milky Way galaxy at this time.* How realistic is this estimate?

If microbial life forms readily, then millions to hundreds of millions of planets in the galaxy have the *potential* for developing advanced life. (We expect that a much higher number will have microbial life.) However, if the advancement to animal-like life requires continental drift, the presence of a large moon, and many of the other rare Earth factors, then it is likely that advanced life is very rare and that Carl Sagan's estimate of a million communicating civilizations is greatly exaggerated. If only one in 1000 Earth-like planets in a habitable zone really evolves as Earth did, then perhaps only a few thousand have advanced life. Although it could be argued that this is too pessimistic, it may also be much too optimistic. Even so, we cannot rule out the possibility that Earth is not unique in the galaxy as an abode of life that has just recently developed primitive technologies for space travel and interplanetary radio communication.

The Rare Earth Equation

Perhaps we can suggest a new equation, which we can call the "Rare Earth Equation," tabulated for our galaxy:

$$N^* \times fp \times ne \times fi \times fc \times fl = N$$

where:

N^* = stars in the Milky Way galaxy
fp = fraction of stars with planets
ne = planets in a star's habitable zone
fi = fraction of habitable planets where life does arise
fc = fraction of planets with life where complex metazoans arise
fl = percentage of a lifetime of a planet that is marked by the presence of complex metazoans

And what if some of the more exotic aspects of Earth's history are required, such as plate tectonics, a large moon, and a critically low number of mass extinctions? When any term of the equation approaches zero, so too does the final result. We will return to this at the end of this essay.

If animal life is so rare, then intelligent animal life must be rarer still. How can we define intelligence? Our favorite definition comes from Christopher McKay of NASA, an astronomer, who defines intelligence as the "ability to construct a radio telescope." Although a chemist might define intelligence as the ability to build a test tube, or an English professor as the ability to write a sonnet, let us for the moment accept McKay's definition and follow the lines of reasoning he sets out in his wonderful essay "Time for Intelligence on Other Planets," published in 1996. Much of the following discussion comes from that source.

McKay points out that if we accept the "Principle of Mediocrity" (also known as the Copernican Principle) that Earth is quite typical and common, it follows that "intelligence has a very high probability of emerging but only after 3.5 billion years of evolution." This supposition is based on a reading of Earth's geological record, which suggests to most authors that evolution has undergone a "steady progressive development of ever more complex and sophisticated forms leading ultimately to human intelligence." Yet McKay notes that evolution on Earth has *not* proceeded in this fashion but rather has been affected by chance events, such as the mass extinctions and continental configurations produced by continental drift. Furthermore, we believe that not only events on Earth, but also the chance fashion in which the solar system was produced, with its characteristic number of planets and planetary positions, may have had a great influence on the history of life here. . . .

McKay [believes] that complex life—and even intelligence—could conceivably arise faster than it did on Earth. If we accept McKay's [theory], a planet could go from an abiotic state to the home of a civilization building radio telescopes in 100 million years, as compared to the nearly 4 billion years it took on Earth. But McKay also concedes that there may be other factors that require a long period:

> What is not known is whether there is some aspect of the bio-geochemical processes on a habitable planet—for example, those dealing with the burial of organic material, the maintenance of habitable temperatures as the stellar luminosity increases gradually over its main sequence lifetime, or global recycling by tectonics—that mandates the long and protracted development of the oxygen-rich biosphere that occurred on Earth. Other important unknowns include the effect of solar system structure on the origin of life and its subsequent evolution to advanced forms.

His inference is that plate tectonics has slowed the rise of oxygen on Earth. But it also may be necessary to ensure a stable oxygenated habitat, just as having the correct types of planets in a solar system is important as well.

In their 1996 essay "Biotically Mediated Surface Cooling and Habitability," Schwartzman and Steven Shore tackle this same problem and reach a different conclusion: They believe that the most critical element in determining the rate at which intelligence can be acquired is a potentially habitable planet's rate of cooling. Their point is that complex life such as animals is extremely temperature-limited, with a very well-defined upper temperature threshold. Although some forms of animal life can exist in temperatures as high as 50°C or sometimes even 60°C, most require lower temperatures, as do the complex plants necessary to underpin animal ecosystems. A maximum temperature of 45°C is probably realistic. It is thus the time necessary for a planet to cool to below this value that is critical, according to these two authors. Many factors affect the time required, including the rate at which a star increases in luminosity through time (which works against cooling), the volcanic outgassing rate (which also works against cooling, because such outgassing puts more greenhouse gases into a planetary atmosphere), the rate at which continental land surface grows (as continents grow, planets usually cool), the weathering rate of land areas, the number of comet or asteroid impacts and their frequency, the size of a star, whether or not plate tectonics exists, the size of the initial planetary oceans, and the history of evolution on the planet.

Earth Is Extraordinarily Rare

With this in mind, let us return to our Rare Earth Equation and flesh it out a bit by adding some of the other factors featured in this essay.

$$N^* \times fp \times fpm \times ne \times ng \times fi \times fc \times fl \times fm \times fj \times fme = N$$

where:

N^* = stars in the Milky Way galaxy
fp = fraction of stars with planets
fpm = fraction of metal-rich planets
ne = planets in a star's habitable zone
ng = stars in a galactic habitable zone
fi = fraction of habitable planets where life does arise
fc = fraction of planets with life where complex metazoans arise
fl = percentage of a lifetime of a planet that is marked by the presence of complex metazoans
fm = fraction of planets with a large moon
fj = fraction of solar systems with Jupiter-sized planets
fme = fraction of planets with a critically low number of mass extinction events

With our added elements, the number of planets with animal life gets even smaller. . . .

Again, *as any term in such an equation approaches zero, so too does the final product.*

How much stock can we put in such a calculation? Clearly, many of these terms are known in only the sketchiest detail. Years from now, after the astrobiology revolution has matured, our understanding of the various factors that have allowed animal life to develop on this planet will be much greater than it is now. Many new factors will be known, and the list of variables involved will undoubtedly be amended. But it is our contention that any strong signal can be perceived even when only sparse data are available. To us, the signal is so strong that even at this time, it appears that Earth indeed may be extraordinarily rare.

THE POSSIBILITY OF COMPLEX EXTRATERRESTRIAL LIFE IS NOT REMOTE

SETI Institute

The SETI (Search for Extraterrestrial Intelligence) Institute conducts scientific research to determine the nature and prevalence of life in the universe. In the following selection, the SETI Institute refutes the contention of Peter Ward and Donald Brownlee that the possibility of intelligent life elsewhere in the universe is remote. According to the institute, factors that are known to contribute toward the creation and maintenance of complex life forms are not as rare as Ward and Brownlee maintain. Furthermore, the institute asserts, none of Earth's properties are unique; therefore, they could feasibly be found on other planets. The institute concludes the only way to determine if intelligent life exists in the universe is to look for it.

Could we be alone in our part of the galaxy, or more dramatic still, could we be the only technological society in the universe? A *New York Times* (2/8/00) science article has reported the opinion that complex life might be decidedly uncommon. The article discusses a new book—*Rare Earth*—in which Peter Ward and Donald Brownlee (University of Washington) point out the many ways in which our solar system may be optimized for our existence. If that's the case, then the other solar systems of the cosmos might support life, but little of it would be more sophisticated than "shower scum."

A Controversial Subject

SETI researchers have long thought about this subject. The idea that intelligent life might be widespread has always been controversial. Many biologists, including Ernst Mayr and Stephen Jay Gould, have argued that humankind's recent appearance on this planet was contingent upon an unlikely chain of events. For example, had the dinosaurs and half of all other species not been wiped out 65 million years ago, we wouldn't be here, stacking their bones in our museums. Similarly, it has been suggested many times that circumstances that seem peculiar

Reprinted, with permission, from "Rare Earth? Are We So Special?" article on the website of the SETI Institute, www.seti-inst.edu/science/rare_earth.html.

to Earth—a large moon, plate tectonics, a salubrious atmosphere, a placid sun, or even the presence of beefy planetary neighbors such as Jupiter and Saturn—were all essential to the evolution of terrestrial intelligence.

But were such things truly "essential," or merely a help? When we tote up a laundry list of Earth's astronomical properties it may seem to imply that our existence is remarkable, and possibly even unique. That could be true, of course. But one should always be leery of probabilities calculated after the fact. A mutt off the street could have a spot on one ear, two eyes of different color, and a limp. In other words, it would be a mutt unlike most other dogs (and if examined carefully enough, a mutt unlike all other dogs). But this hound could also have fleas. Would we be correct in concluding that dog, so obviously special, is the only one likely to have fleas?

The question is not how idiosyncratic is Earth, but whether our world enjoys circumstances that are simultaneously rare and essential to complex life. As far as we know, this isn't the case.

Consider the Sun

Consider our stellar host. Is the Sun a rarity in the galaxy? Sol is a G-type star of medium brightness, endowed with long life and a tranquil disposition. Sure, it spews a moderate amount of ultraviolet our way, but most of this dangerous radiation is screened from Earth's surface by atmospheric ozone. Nonetheless, we know that most other stars are, indeed, unlike the Sun. Some are bigger and brighter, and a great deal more are runty and dim.

The brighter stars die young, typically in only a few dozen million years, hardly sufficient to incubate complex life. The plentiful dimmer stars are both more volatile (they routinely flare) and more likely to trap habitable planets in a synchronous rotation, with one side always turned toward their stellar masters. Unless such a planet has a thick mantle of air, this could wreck the weather (although some recent research indicates that even a somewhat thinner atmosphere laced with greenhouse gasses would make these tidally tugged worlds habitable). However, if we wish to be conservative, we might opine that dim stars are poor bets for supporting sophisticated biology.

Large stars are non-starters, and small stars are problematic. But sun-like stars are not rare. Approximately one star in ten is similar to Sol, and that means that tens of billions of them inhabit our galaxy. Indeed, even stars somewhat brighter than the Sun, which because of their heightened ultraviolet production might be thought dangerous, could be the loci of life. Jim Kasting has shown that the enhanced ultraviolet radiation from F-type stars would produce so much atmospheric ozone that planetary surfaces could be very well shielded.

Rare Earth's authors suggest that the composition of our solar system, including the materials necessary for making rocky planets, might be unusual. The chemical composition of stars certainly varies, and some are poorly endowed with heavy elements (everything more massive than helium), essential ingredients for the formation of both planets and protoplasm. This is described by astronomers as a variation in "metallicity." Globular clusters, composed of the galaxy's oldest (and most heavy-element poor) stars, have metallicities of 0.1% or more, or about an order of magnitude less than the Sun (this sprinkling of heavy elements is larded into the interstellar medium by supernova explosions). But models of planet formation don't rule out the formation of new worlds even in these depleted systems. And there's still plenty of raw material. Note that the mass of the Earth is only about 0.0003% that of the Sun, so even in the metal-deprived neighborhoods of globular clusters there is more than enough suitable material for constructing Earth-like planets.

In addition, the overwhelming majority of luminous stars are in the galaxy's disk, not in the globulars. Since star formation has remained relatively constant since the Milky Way's birth, roughly one-half of all stars have metallicities comparable to, or even greater than, that of the Sun. The composition of our solar system will be similar to that of billions of other sun-centered solar systems. Roughly half of these will be old enough to have incubated intelligent beings.

Other Factors

But what about such possible terrestrial peculiarities as plate tectonics or the companionship of a large moon? Plate tectonics are useful for cycling carbon between the atmosphere, the ocean, carbonate rocks, and living things. This happens because the slip-slide of the oceans pushes carbonate rocks on the sea bottom (such as limestone) under the continental margins. The carbonates melt and are blown back into the atmosphere by volcanoes. But tectonic activity needn't be either rare or short-lived. It is driven by internal heat produced by radioactive materials in Earth's center. Other planets of similar composition and of a size at least that of Earth will also enjoy a dynamic surface. Indeed, if the positions of Venus and Mars had been interchanged at birth, then Venus might be both tectonically and biologically active today. There is nothing miraculous about tectonic activity, and in fact there is some evidence that it has occurred on both Mars and Venus.

Does sophisticated life require a large moon? Some have argued that it does, for the moon helps stabilize Earth's rotation axis and keeps it from possibly dangerous tilts. Other Earth-like planets might not be blessed with such a hefty moon, as it seems that our own natural satellite was produced in an accidental collision between Earth and a Mars-sized or larger asteroid more than 4 billion years ago. But computations have shown that if the moon had not been formed, our planet would spin faster—fast enough, in fact, to stabilize it against major tipping. In addition, even if an Earth-like planet occasionally does a spin flip, it will spend 10 million years or more doing so. Life can probably adapt to such slow changes. Indeed, it already has, during episodes of polar wander on Earth.

Finally, there's the matter of biological evolution to intelligence. Is there some good reason why evolution should occasionally produce sentient beings? The answer is, and remains, controversial. However, the fossil record tells us that the maximum degree of encephalization—which (crudely speaking) measures the ratio of brain to body mass—has increased considerably in complex animals for the last 100 million years or so. The dinosaurs were bulky but not bright. Even the most cerebral of these lumbering lizards had less brain power, as judged by encephalization, than an ostrich. The mutual stimulus to increased mental capability inherent in predator-prey activity has led to increased encephalization for some segments of the animal kingdom. Consequently, many of today's animals would handily outscore their Mesozoic predecessors on any IQ test. Human-level sophistication could be a common outcome of this ratcheting up of neural capability, although it is the uncertainty of this conclusion that drives us to look for intelligence elsewhere.

No Unique Properties

The bottom line is this: our solar system and our Earth have "personality"—they exhibit properties that might be found only occasionally in other star systems. But there are no show stoppers that mandate against the evolution of sophisticated life elsewhere. Our solar system has no properties that are obviously essential to complex or intelligent life that other worlds would never have. We may even have been cheated out of some helpful phenomena that could have sped evolution on Earth. We might be less lucky than we recognize, and creatures on other worlds may regard with disappointment the nature of our planet.

Possibly we have failed to appreciate the importance of some aspect of our situation, and the fact that we are functionally very special—that intelligence and complexity are a rare exception to the cosmic rule. But while no experiment can prove we're alone, SETI experiments could show that we're not. Consequently, arguments about whether intelligence is a rare event in the universe—while instructive and interesting—may be akin to discussing with Columbus whether there's any chance his voyage will uncover a new continent. Discussion does not settle such arguments. Only experimentation can.

We should continue to sail the seas of discovery, for only by doing so do we have a chance of revealing our place in the cosmos. As Giuseppi Cocconi and Philip Morrison stated four decades ago, "the probability of success is difficult to estimate, but if we never search, the chance of success is zero."

THE FACE ON MARS MAY BE AN ARTIFICIAL FORMATION

Mark J. Carlotto

Satellite photographs from 1976 of the Martian surface show a formation that has the characteristics of a human face. While some claim that the apparent Face on Mars is merely the result of erosion and the tricks of light and shadow, others, including author Mark J. Carlotto, maintain that the face may be an artificial formation that was built for some unknown reason by intelligent life forms eons ago. Carlotto points out that several other landforms, many of which are geometric and appear to be in a pattern, are located near the face. It is extremely unlikely, Carlotto contends, for all these objects to have been carved over the ages by random forces of nature. Carlotto has twenty years of experience in digital image processing, enhancement, and restoration. He has written several articles about the Face on Mars and is the author of *The Martian Enigmas: A Closer Look,* from which this selection is excerpted.

On the surface of Mars lies a formation that looks remarkably like a humanoid face. Forever staring up into the vastness of space it has attracted our attention. For some, that is why it is there, beaconing us to come and explore. Others believe that it is simply an odd looking geological landform—a formation carved over the ages by the random forces of nature. That it is our imagination and our need to find other life in the universe that makes us see it as an intelligently crafted object.

It Is Not Alone

And perhaps this is all that could be said of the Face on Mars. Provided it was alone. But it is not alone. Nearby are other strange looking objects. Some quite geometrical in shape. A number of them look like pyramids, one apparently five-sided. Moreover the objects seem to be arranged on the Martian surface in an organized pattern. Again, maybe it is our imagination tricking us into seeing something that is not there. But there is more.

Looking closer we see that there are subtle details within the Face as well as several of the other objects. Details that should not be there if these objects are natural. Logically these details should have been obliterated by erosional processes rather than preserved.

But once again we stop ourselves and ask: Could all that we see be image or processing artifacts? Questions that need to be asked whenever one interprets digital satellite imagery. And once again the answer seems to be no. What we see is present in more than one image. In fact all of these objects were imaged at least twice—about 35 days apart under slightly different sun angles.

If the Face on Mars and nearby pyramidal objects are artificial, they beg us to compare them with the Egyptian Sphinx and pyramids. But there is no comparison. These objects on Mars are enormous even by the standards of those that built the pyramids on the Giza plateau. The Face is over a mile long and is three times taller than the Great Pyramid. Where the Great Pyramid has four sides, each about 687 feet long, this apparently five-sided pyramid on Mars has sides well over a mile long. The pyramids on Mars are roughly 100 times larger in area and 1000 times greater in volume than the Great Pyramid, one of the largest structures on Earth!

The enormity of scale of these objects on Mars would seem to imply that they cannot be real. They must be natural geological formations. But then again, since the gravity on Mars is about one third that of Earth, one, with the same technology, could build structures considerably larger. But then again we are at a loss to explain them because we cannot even explain how the Egyptian pyramids were built let alone scores of other terrestrial enigmas.

So we seem to have a real mystery on our hands.

An Important Discovery

The possible detection of artificial structures on another planet would be among the most important discoveries in the history of mankind. It should have triggered a major revolution in scientific, social, and philosophical thought. But it did not. The best minds on Earth should be trying to figure out why it is there and be working toward a mission to Mars to find out. But they are not.

Perhaps we have not been ready. For up until recently even the possibility of finding microbes on Mars was considered remote. But with the 1996 discovery of what appear to be fossilized micro-organisms in a meteorite thought to be from Mars, scientific attitudes are changing. And the public seems hungry for more. Visionaries are beginning to talk of Mars as the next frontier. Engineers have developed low-cost solutions to get us there. Scientists are even talking about terraforming Mars—transforming it into another Earth.

Meanwhile the Face on Mars waits patiently for us. It has waited for millennia. It can wait a little longer.

Return of the Martian Canal Builders?

Our interest in Mars is a relatively recent phenomena. It began late in the 18th century as gradually improving telescopes showed Mars to be like Earth in some ways. This touched off speculation about life on Mars which grew in the 19th century and led to the great Martian canal controversy. And almost a century later the canals still haunt us.

Shortly after the Face was first imaged in the summer of 1976, the noted geologist Harold Masursky quipped, "This is the guy that built all of [Percival] Lowell's canals." Most planetary scientists seem to agree with Hal Masursky, Carl Sagan, and others who say the Face on Mars, like the canals, is an illusion. According to Sagan, "Lowell always said the regularity of the canals was an unmistakable sign that they were of intelligent origin. This is certainly true. The only unresolved question was which side of the telescope the intelligence was on."

While some planetary scientists are open to the possibility that the Face and nearby objects may be artificial—that they deserve a closer look, most either do not, or will not see the Face as anything other than a strange looking rock formation. While this may seem strange, it is, in fact, to be expected. According to Thomas Kuhn in his book the *Structure of Scientific Revolutions,* normal science is based on paradigms, ways of looking at the world. "No part of the aim of normal science is to call forth new sorts of phenomena; indeed those that will not fit into the 'box' [that the paradigm supplies] are often not seen at all."

A large fraction of the planetary science community is made up of geologists. And geologists see Martian surface features simply as geology. They do not see the possibility that the Face might be artificial because they can not. It lies outside of their discipline.

Why a Human Face?

Other scientists refuse to see the Face as an artificial object because its existence would undermine their world view. If it is artificial then either it was built by a technological civilization native to the planet, by visitors from outside of our solar system, or by a previous technological civilization from Earth.

So they ask, how can there be a human face on Mars? At best Mars was wet and warm for its first billion years. It took five billion years for life to start and evolve on Earth to its present form. How could the same thing have happened on Mars in only one billion years?

Or they ask, if it was constructed by visitors from outside of our solar system, why did they pick a human face? Were they humanoid? Whether they were Martians or visitors from outside of our solar system, the human form creates another problem. According to [Charles] Darwin, evolution does not follow a predefined path. It seeks no goal.

Perhaps it is even random. So how can the same form evolve on two different planets, each with radically different histories?

Or, they ask, how could it have been built by an earlier race of humans from Earth. They speculate that it is millions, perhaps billions, of years old. But even if it is only tens of thousands of years old, how could it have been built by men who were just discovering stone tools at the time? They say that there is no evidence of an earlier advanced technological civilization on Earth. Yet there are many ancient mysteries that we still cannot explain.

A Major Paradigm Shift

If the Face on Mars is artificial it will trigger a major paradigm shift. And science has historically resisted such shifts. This too must be factored into what is happening. Could the desire to protect the paradigm cause some scientists to be a little too zealous? To bend the rules of the game a little? Not merely to be skeptical but to go on the offensive and try to discredit those who believe the Face may be artificial? To distort the truth a little? Never to mention certain researchers by name—researchers who have actually subjected their work to peer-review and have had it published in technical and scientific journals? Research that presents strong evidence about not just one isolated anomaly, but a collection of unusual formations. The diversity of which is hard to explain.

Or could the desire to protect the paradigm cause other scientists to go into a state of denial, to say the Face can't be there so it isn't? To make the whole matter into a joke?

I raise these questions because many people look to scientists for answers. What should I think about this face on Mars? Could it be real? What does it mean? If science ridicules the Face on Mars, so will the people who rely on these scientists to help them sort out the facts. They too will think that it is just a joke. Undoubtedly they will be entertained by tabloid accounts of ruined temples on Mars, the voice of Elvis Presley emanating from the Face on Mars, and other such nonsense. But ultimately they may be misled.

The Evidence

So what evidence does exist to support the claim that the Face and other nearby objects on Mars are artificial in origin? The late astronomer Carl Sagan's famous quote, "Extraordinary claims require extraordinary evidence," places the responsibility for providing the evidence on the individual or group which makes the claim. Most will admit that there is no single piece of direct evidence, no "smoking gun," to support the claim for artificiality. But what about circumstantial evidence? Evidence that can have other interpretations. So how much evidence is there? Just a few interesting coincidences, or is there more?

The recent announcement concerning the possible discovery of fossilized micro-organisms in a meteorite has sparked tremendous interest in Mars exploration. This is certainly an extraordinary claim. But is the evidence that extraordinary? Some scientists say no. Even the researchers admit that each individual piece of evidence has other interpretations, and that their conclusion is based on not one piece of direct evidence (a live microbe) but on several pieces of indirect evidence pointing to the same conclusion. It is the convergence of circumstantial evidence that leads them to conclude that they have found a fossilized microbe.

But the point here is not whether the evidence is strong or weak but rather that the question of microbial life is being debated at all. In other words, why is the search for microbes more scientific than that for artificial structures? Perhaps because finding microbes is less likely to upset the paradigm than the discovery of a mile long humanoid face on Mars.

Since the early 1960s, radio-telescopes have been used to search for signals from outside of our solar system. The search for extraterrestrial intelligence (SETI) is predicated on the assumption that there are a sufficient number of advanced technological civilizations in the galaxy to warrant such a search in the first place. Could a small number of these technological civilizations have developed the technology for space travel, and one of them, in the distant past, visited our solar system, perhaps Mars, and left artifacts on its surface? One would think that if the search for extraterrestrial radio signals is a legitimate scientific endeavor, why isn't the search for extraterrestrial artifacts? Again, could it be that the discovery of radio signals from a star system many light years away would be much less threatening than the discovery of real physical artifacts practically next door to us on Mars?

The Theory Can Be Tested

So aside from the possibility that the Face on Mars could upset many of our treasured paradigms, why isn't the study of the Face scientific? Clearly the phenomena itself is within the realm of science. The hypothesis that the Face is artificial is falsifiable. It can be tested.

Some scientists say the imagery is too blurry, not detailed enough to support the claim for artificiality at this time. That it is only through a magical thing known as "image processing" that one can be made to see. And what one then sees is not really there. It is noise in imagery enhanced through the improper use of image processing techniques. But is that really the case? Have so many people been simply the victims of "wishful seeing," of having their perceptions biased by their expectations, by what they want to see?

Perhaps it is the methods that have been used or the reasoning of the researchers that is unscientific. But what if the methods have actually been tested? For example, what does one make of a mathematical technique which detects manmade objects in terrestrial satellite imagery that happens to find that the Face is the least natural object within this part of Cydonia [the name of the region where the Face is located]? Or the results of a three-dimensional analysis of the Face that shows that it retains its appearance over a wide range of lighting conditions and perspectives? In other words, that it is not an optical illusion?

The Face on Mars reminds some scientists of the great controversy over the canals on Mars around the turn of the [twentieth] century. With gradually improving telescopes, astronomers were beginning to see indications of subtle Martian surface features for the first time. For many, their eyes integrated these subtle patterns into linear features. Some even interpreted these lines to be waterways built by Martians to distribute water over their dried out and dying planet. But then as the telescope continued to improve, astronomers began to realize that the canals were an illusion after all. Many of today's scientists believe that this is exactly what is happening when it comes to the Face on Mars. That higher resolution images will prove the Face and other objects to be natural rock formations.

Perhaps. But opinions about the possibility of life on Mars have changed dramatically over the years—from Percival Lowell's canals, to the dead planet imaged by the early *Mariner* probes, to the enormous volcanoes, great canyon systems, and channels carved by water discovered by *Mariner 9* and later confirmed by *Viking*.

One lesson is clear, particularly when it come to Mars—be prepared for the unexpected.

It Is Unlikely That the Face on Mars Is an Artificial Formation

Shaun Cronin and David Vaughan

In the 1970s, a U.S. space probe circling Mars photographed a rock formation that looks like a human face. Many people believe that the Face on Mars was built eons ago by some form of intelligent life. In the following essay, Shaun Cronin and David Vaughan explain why they are skeptical of such a theory. According to the authors, there is no strong scientific evidence supporting the belief that aliens built the face. Furthermore, they maintain, using a photo of a rock that happens to look like a human face as the sole proof of the contention that there is intelligent extraterrestrial life is illogical and pseudoscientific. They conclude that the theory is popular because it comforts those who need to believe that humans are not alone in the universe. Cronin and Vaughan operate a website, www.skepticalskoundrels.com, in which they analyze many unusual and controversial events and phenomena.

What is the hypothesis [about the Face on Mars] and who are the general supporters?

The main proponent is Richard Hoagland. His website [http://www. enterprisemission.com] is devoted to "research" concerning artificial structures on Mars. Other than faces, cities and pyramids, Hoagland has claimed to have found howitzers, lawn furniture and even Nazi insignia on the Pathfinder probe. To that end, Hoagland has written books, released videos and made many a speech testifying to the "reality" of artificial structures on Mars.

There are a number of researchers who are taking a more cautious and less sensationalistic approach to the subject. Mark Carlotto is a good example. While Carlotto seems to cautiously support the idea that artificial structures exist on Mars, he is rather more restrained than Hoagland. His website is at http://www.psrw.com/~markc/marshome. html.

Of course, do a search on the Web and you will find many a site devoted to the possibility of artificial structures on Mars. These pages range from the marginally amusing to the downright ridiculous. The

Reprinted, with permission, from "Face on Mars FAQ," by Shaun Cronin and David Vaughan, published at www.skepticalskoundrels.com/marsques. html.

Mars Page O' Links [www.skepticalskoundrels.com/marslinks.html] has some web sites ranging from the skeptical to the credulous to the "we are really not sure."

The Belief in the Face on Mars

Why do these supporters think there is a face on Mars?

The belief in a face on Mars is related to why people believe that aliens regularly visit Earth, that humans are a genetic experiment conducted by aliens and a myriad of other paranormal beliefs. These beliefs are manifestations of our hopes and desires. The seductive nature of such claims appeal to the basic question of "are we alone in the Universe?"

Hoagland and his cronies would answer "No! We are not alone. We have a message from aliens!" Though the notion that we are may frighten people, the truth or falsity of a Martian relic is still debatable, and there is no credible evidence to support it. If you are one of those taken to fear of being alone in the universe, take heart! The odds that other intelligent life exists elsewhere seems to all but prove we are not.

The history of the mythological romance that Homo sapiens have had with Mars makes the idea appealing. The idea of a humanoid face peering into space on the surface of Mars appeals to our anthropocentric notions of what purpose aliens may have had or what they may have looked like.

Some supporters cling to the idea of structures on Mars because they have invested too much into trying to shore up the leaky ship that is their belief. A parallel can be seen with those that still cling to the belief that an alien spacecraft crashed at Roswell in 1947. More than one UFO researcher has shown that the Roswell story is full of holes yet a few prominent "researchers" do exist. To concede that Roswell has become a joke is something that they cannot accept. The more time, money and prestige one has developed into formulating an idea may lead to a reluctance to abandon the idea even if the supporting evidence is highly dubious.

Are there any other supposedly intelligently designed monuments on Mars?

The same area of Cydonia is claimed to also show evidence of pyramids and a city. Some researchers have claimed other areas show signs of habitations. A few people have stared at the Pathfinder photographs way too long and claim to have found life on Mars. Browsing through the Mars Page O' Links will satisfy your curiosity.

Why do you bother to argue about this face? Does it hurt anyone to believe there is a face on Mars?

There seems to be a trend towards the idea that all beliefs are valid. Indeed, in some postmodernistic circles this is actively preached. There is no problem with people believing what they want. However, when they want their beliefs to be presented as valid based on shoddy evidence, then there is a problem. Critical thinking seems to be at a premium these days. As a consequence, the "Face on Mars" proponents make money off the gullible. They make outrageous claims that there was an alien civilization on Mars and charge accordingly for releasing that knowledge.

Even accepting that there isn't a face or other monument of intelligent design on Mars, hasn't NASA often lied about what they find and what they do?

No one has ever presented any credible evidence that NASA has lied to the public concerning the Face on Mars. The myth that NASA has lied to the public concerning what they have found on Mars is an extension of the age-old motif of the UFO mythos of a cover-up. The accusations of cover-up lie in the false premise that if NASA photographs show no evidence of artificial structures on Mars then NASA is lying. The root of this is the unfounded belief that such structures do exist. It is quite common for proponents of paranormal claims to dismiss contradictory evidence to their claims as being part of a government conspiracy to hide the truth.

"Disinformation"

An extension of this is the current UFOlogy buzzword, "disinformation." The more the word "disinformation" is used the more it seems to be a synonym for "gullible." Every time a hoax or evidence contrary to the beliefs of the True Believers is revealed, those who make such revelations are applied with a brush that paints them as a disinformationist (or whatever). This then allows those that have been fooled to avoid the problems of their credulity. The connotation behind disinformation is that there is a conspiracy to discredit the seekers of the Truth. Unfortunately, what these seekers of the Truth fail to realize is that they themselves do more damage to their credibility than any disinformation campaign.

"Disinformation" is becoming a far too convenient word that allows those that should know better to escape responsibility, whether it be personal or to a larger community.

Some Martian monument proponents claim that what they do is science because they use mathematical calculations to show that the structures may be artificial. Why do skeptics dismiss such seemingly legitimate attempts at gathering evidence regarding the Martian structures?

For very simple reasons, actually. First of all the math used would *have* to be arbitrary, as the only thing they have to base their math on is a two-dimensional representation of a formation on Mars. How are we, or they, for that matter, to know that these figures they use are accurate? Well, simply put, we can't. The entire exercise has to begin with certain assumptions, and a great deal of faith in their ability to make those assumptions. Now, let's consider the consequences of their math. Mars had intelligent life in the not-so-distant past that made monuments of human figures and symbols. Is there any evidence of this besides math based on a two-dimensional photo taken from a great distance? Obviously not. Is it therefore reasonable to assume that the math is correct and *every other hypothesis* about the formations on Mars is wrong? Of course not. You have to remember that to grant that there are "faces" on Mars is not a single statement. It is manifold in its influence. It negates, simultaneously, *all* other accepted facts about Mars. The amount of information that would have to be discarded prevents any real consideration of the theory of intelligent life on Mars sculpting faces. That is why skeptics dismiss these attempts, which, apart from being legitimate, are fanciful and useless to *scientific* discourse, except, perhaps, the discussion of brain function and its ability to recognize patterns in many chaotic jumbles of visual information. Like clouds, for instance. We all see shapes there, but few of us would create invisible beings shaping clouds into terrestrial items.

No Corroborating Evidence

No one has ever set foot on Mars. How can one be skeptical of that artificiality of the structures when they haven't even been to Mars?

There is no corroborating evidence at all to support the hypothesis of artificial structures on Mars. Look at it this way: given that there has *never* been *any* substantiated evidence of *any* life on Mars, and that the only evidence is a few photos, which is more likely—the overwhelming observations of a lifeless planet and explanations of formations from natural phenomenon, *or* that someone saw a picture of a rock that looked like a human face and made a mistaken assumption that it was artificially produced?

Some of the claims made by certain proponents appear to be quite outrageous. Should we dismiss all such claims made by these people or just the ones that refer to howitzers, Nazi insignia and so on?

I see no reason to differentiate between any of them. It is no more outrageous that there is a howitzer on Mars than that there is a monument designed to look like a terrestrial human. Once you enter the level of bizarre there are few intermediate steps.

There has been a meteorite that was discovered with what appeared to be Martian fossils as well as evidence that water once existed on Mars. This points to evidence that life once existed on Mars. If so, why are skeptics so keen to dismiss such claims that there was an ancient civilization on Mars?

Prehistoric bacterial life does not, in any way, translate into inevitable intelligent life. That very important point notwithstanding, there is also the confidence one assumes for this rock purporting to show life from Mars. Without getting too technical, there are some very real reasons to be cautious of the claims made for this rock. *Scientific American* and *The Planetary Society* update the research fairly regularly, and it looks like the preponderance of the investigation points to contamination by terrestrial biologicals. However, the point is a good one to make because it delineates pseudo-science from real science. The Mars rock, as it is commonly referred to, was given to others to study and investigate. Lively debate among peers furthers the knowledge of what may or may not be found in this relic. The investigation of the "Face" is impervious to such investigation. The basic argument is, "*If* this looks like a face, then it *is* a face." That is at best skewed logic and not in the least scientific, nor even very rational, but at the same time it is beyond investigation. I can look at the picture and give 20 or 30 alternative explanations, but the idea that possibility equals reality prevents any real counter observation validity. This is a sure sign of pseudo-science. Beware such arguments.

CHAPTER 2

THE SEARCH FOR EXTRATERRESTRIAL INTELLIGENCE

Contemporary Issues
Companion

LISTENING FOR EXTRATERRESTRIALS

William J. Broad

William J. Broad is a science writer for the *New York Times*. In the following reading, Broad discusses how scientists involved in the search for extraterrestrial intelligence (SETI) use telescopes and dish antennae to search for radio signals emanating from space. Although astronomers and scientists have been searching for extraterrestrial intelligence among the stars for decades, he writes, no evidence has been found of its existence. Nevertheless, Broad notes that SETI researchers remain confident that someday they will make contact with aliens, especially since new strategies and ideas are being developed that will allow the searchers to widen the scope of the hunt.

"Sorry guys," Jill C. Tarter said, cutting off chitchat around the control panel, eager to hunt for alien civilizations. "We have the telescope."

Outside in the fading light, surrounded by dense jungle, the receiving dome on the world's largest radiotelescope wheeled into position. It fixed on Barnard's star, a scant six light-years away, the closest star to Earth in the Northern Hemisphere. Closeness meant it had been searched before. But now, the great dish antenna of the Arecibo observatory began to gather in a riot of faint signals, giving the star its most discriminating look yet for hints of invisible planets and intelligent life.

Tense with concentration, Dr. Tarter, 54, closely examined the colored spikes that slowly materialized on her monitor. Each was a candidate, a possible hello from afar. But in the next hour and a half, she, a colleague and a nearby supercomputer rejected them, one by one. The signals turned out to be cosmic static and earthly interference.

She showed no sign of frustration—no sigh, joke or frown. A true believer, she just plowed ahead to examine a night full of stars, sure that some day there would be proof that humans were not alone in the universe.

At a Turning Point

The search for extraterrestrial intelligence, or SETI, is at a turning point after nearly four decades of hard work. With the arrival here of Dr.

Tarter and her crew from the SETI Institute in Mountain View, California, the field has reached a high point in terms of telescope sensitivity, a top goal of the alien hunters.

But it has found no extraterrestrials so far, despite forecasts that they should have been discovered by now. Instead, it has probed the heavens with regularity and heard nothing but a dead silence.

While pressing the hunt here at Arecibo, Dr. Tarter and her peers around the globe are engaged in quiet debate and soul searching over how to proceed. New ideas and strategies are being weighed, including expansions that would enhance telescope sensitivity and widen the hunt to more stars—perhaps 100,000 rather than the 1,000 now targeted.

"When you get far into a project and haven't found what you're looking for, it gives you pause," Seth Shostak, a SETI Institute scientist, said as he trudged down a hillside into the lush bowl that holds the telescope's huge receiving dish, the size of 26 football fields.

Does the silence bother him?

"It's a funny thing," Dr. Shostak replied. "You'd think it gets old. But it doesn't. The equipment keeps getting better and the odds keep getting better, which never happens in Vegas. Also, there are always new ideas. This is a field with new papers, new meetings, new people—which is remarkable considering there is no data. So for me, it's not discouraging at all."

Close up, the great antenna was a grayish sea of thousands of perforated aluminum panels. High overhead, a cable car hauled engineers to the central receiving dome to do maintenance. Dr. Shostak strode under the dish into dim sunlight and surprisingly dense foliage.

Any boa constrictors down there?

"The worst thing is the mud," he called back confidently.

Just a Matter of Time

Like most SETI enthusiasts, Dr. Shostak and Dr. Tarter believe it is just a matter of time before earthlings use such antennas to make contact with aliens. Their faith is rooted in numbers, big ones. The Milky Way is estimated to have 400 billion stars, including the Sun. SETI scientists believe that many of these stars have planets orbiting them as well as advanced forms of life—an idea skeptics deride.

Alien civilizations in the galaxy are likely to number anywhere from 10,000 to one million, SETI enthusiasts say. If the higher density is right, that means advanced beings would inhabit about one in every 400,000 stars. The implication is that even a slow, detailed, comprehensive search from Earth would be mostly a wasteland of late nights, false leads and frustration.

"It may look empty, but it's not," Dr. Shostak said as he sipped a root beer in the observatory's cafeteria, eager to cool off from the wet heat outside.

"O.K., maybe this is a hopeless task, maybe it's impossible," he conceded. "On the other hand, maybe it's like discussing the possibility of whether there's a continent between Europe and Asia in the cafes of Segovia in the 1400's. Until you do the experiment, you don't know."

SETI's Beginnings

The first SETI hunt began in humble circumstances in the Allegheny Mountains of West Virginia. In 1960, Dr. Frank D. Drake, a young scientist at the National Radio Astronomy Observatory there, used an 85-foot antenna to listen around a few stars for alien transmissions. Ever since, a main SETI strategy has been to wield increasingly big radiotelescopes, their size allowing them to gather increasingly faint signals. The dish antenna at Arecibo, 1,000 feet wide, is the biggest of them all.

SETI work started here with a bang. In 1974, at the urging of Dr. Drake, who then ran the observatory, the newly upgraded dish at Arecibo was used to beam a powerful, three-minute message at M13, a dense cluster of hundreds of thousands of stars orbiting the Milky Way. The message was a simple graphic showing the telescope as well as facts about the solar system and humans. The message is still zooming outward. At the speed of light, it will reach M13 in about 21,000 years. A reply from any aliens in that neighborhood would presumably take a similarly long time.

Dr. Tarter got hooked on the field in the mid-1970's, soon after the message was sent. She was then a young astronomer at the University of California at Berkeley. After reading a SETI report, she teamed up with another astronomer to hunt for aliens with the university's 85-foot telescope.

"It was brave of Jill," Dr. Drake said in his autobiography, noting that SETI work back then could hurt a developing career.

By 1985, Dr. Tarter was a senior SETI scientist at the National Aeronautics and Space Administration. She and her colleagues built powerful computers to sift through cosmic and earthly interference and lined up many radiotelescopes, including Arecibo. Their goal was to survey 1,000 nearby stars, all within 200 light-years of Earth.

In 1992, the big search was ready to start. And Dr. Drake, the SETI pioneer, gave it a drum roll in his autobiography, *Is Anyone Out There?* (Delacorte Press, 1992, written with Dava Sobel), saying the find of all time was imminent.

"This discovery," he wrote, "which I fully expect to witness before the year 2000, will profoundly change the world."

But the plug was pulled suddenly in 1993 when Congress decided that SETI was a waste of public money.

SETI Goes Private

Undeterred, Dr. Drake and Dr. Tarter took the hunt private at the SETI Institute. They won substantial support from a handful of silicon moguls, capitalized on the $58 million the Government had invested in gear and forged ahead with the planned search. In 1995, they began roving from telescope to telescope.

Dr. Tarter, who heads Project Phoenix, as the hunt is known, is famous for her zeal. She was the role model for Ellie, the heroine of the 1997 movie *Contact,* based on Carl Sagan's novel, starring Jodie Foster. The two met when it was filmed at Arecibo.

"She's a fantastic person," Dr. Tarter said of Ms. Foster. "She wanted to know what astronomers are like, and do they have big egos." She laughed.

Wearing sandals, her hair tied in a ponytail, Dr. Tarter along with her institute colleagues started observing here September 9, 1998. It was the team's first return since the Federal plug was pulled. A new agreement gives them 2,000 hours of observing time, or about a half-year's worth of 12-hour night shifts, which they plan to spread over the next few years.

A big lure is the new sensitivity of the telescope at Arecibo, which is run by Cornell University in cooperation with the National Science Foundation. In 1997, work here was finished on a five-year, $27 million face lift that quadrupled the telescope's ability to pull in faint signals.

The Hunt

As always, the hunt focuses on close stars, since signals from their inhabited planets would be strongest. And it concentrates on ones similar to the Sun, the only star known to support life. Lastly, it tends to search older stars, since it assumes advanced life takes time to evolve.

At Arecibo, candidate alien signals are compared with readings from a radiotelescope nearly halfway around the world at Jodrell Bank in Britain. The comparison helps identify and rule out local earthly interference, which is exploding with the rise in satellites and cell phones.

"That is some hellacious thing," Dr. Tarter said as she glared at an interference spike. "It produces system indigestion."

Around midnight on September 15, 1998, Dr. Tarter and Dr. Shostak were searching for aliens around EQ Pegasi, an unremarkable star 21 light-years away. Suddenly, the team's automatic search program started moving the telescope off the star, seeking to find out if a strong incoming signal was simply interference.

The signal died away, as it would if it originated from the star. And it returned when the telescope refocused on EQ Pegasi.

Rising out of their chairs, electrified by the drama, the two astronomers studied the signal's high rate of drift. That suggested the transmitter was based on either a spinning satellite or a rapidly turning planet.

"Had Jill and I stared any harder at that display screen, we would have bored holes in the phosphor," Dr. Shostak recalled.

Again, the telescope was moved off target as the computer sought to double-check the signal's place of origin. This time the signal stayed on, meaning it was from a satellite.

"It was a big disappointment," Dr. Shostak admitted the next day. "I thought, 'Hey, this is the big one.'" Only a few times before had the team had such a sense of being on the verge of discovery.

Days later, on September 21, Hurricane Georges slammed into the area, uprooting trees, hurling debris that damaged some antenna panels and temporarily disrupting the alien hunt.

New Strategies

Their faith apparently intact, despite the long hours and decades of failure, the alien hunters are planning in their spare time a new generation of strategies and gear. A committee based at the SETI Institute, including Dr. Tarter, Dr. Shostak and 31 other experts, began meeting in 1997 to chart a path into the future.

At the controls of the Arecibo search, Dr. Tarter became quite animated as she described futuristic arrays of hundreds and perhaps even thousands of small dish antennas tied to one another electronically, scanning the sky for the advanced civilizations she knows are out there.

"We have to grow this," she said of untried gear that one day might dwarf Arecibo in sensitivity. "You have to crawl before you walk."

Her ultimate dream is to build an observatory on the far side of the Moon, free of earthly interference, scanning the heavens for an unfamiliar hello. She wants to be there herself, at the controls.

WHAT HAPPENS IF SCIENTISTS FIND EXTRATERRESTRIALS?

Seth Shostak

Seth Shostak, the author of *Sharing the Universe: Perspectives on Extraterrestrial Life,* is a radio astronomer who uses radio telescopes to search for extraterrestrial intelligence for the SETI Institute's Project Phoenix. In the following selection, Shostak discusses the ongoing search for extraterrestrial intelligence and the possible results if that search succeeds. Shostak maintains that the discovery of an alien signal would have an enormous impact on the world as people come to the realization that Earth is not unique. In addition, he notes, contact with an extraterrestrial culture would have long-term consequences on the human race and psyche. Although Shostak admits that the full impact of such a discovery is difficult to predict, he believes that fears of hostile aliens are unwarranted.

The scientific discovery of the millennium: That's how scientists pursuing the search for extraterrestrial intelligence describe the possible consequence of their efforts. Is their prediction hyperbole or understatement? Would detection of a radio transmission from a distant star system truly revolutionize the world?

SETI programs are sometimes compared to Columbus's first voyage. But the analogy is imperfect. His discovery was momentous, but unexpected. In contrast, the discovery sought by SETI researchers has been clearly defined in advance. If, as most scientists believe, the detection of an alien radio signal would have pervasive and long-lasting consequences, it makes sense to be prepared. Astronomers, anthropologists, historians, and science-fiction writers have all given thought to what may happen the day after our radio telescopes intercept the faint tones of an extraterrestrial transmission.

In some ways, the immediate effect of a detection on scientific understanding might not be overwhelming. If our telescopes merely succeeded in picking up a narrow-band whine from space, a carrier without a message, then humanity's newfound knowledge would consist of a few facts about an alien planet's dynamics.

Reprinted from "The Day After," by Seth Shostak, *Mercury*, March/April 1996, by permission of the Astronomical Society of the Pacific, San Francisco.

But while the scientific impact of discovering a distant intelligence is indeterminate, the conceptual impact would surely be mind-boggling. We would be confronted with the proof, rather than simply the possibility, that we are not the top dog of the Galaxy. At a minimum, such a discovery would likely force an unprecedented humbling of *Homo sapiens*. Depending on your point of view, this sudden demotion would lead to either mass hysteria, government cover-ups, or new feelings of peace and community. The long-term effects might be more dramatic yet.

Sounding the Alarm

The drama would begin at an observatory. This is where sophisticated receivers sift through the pervasive hiss of natural static in search of the narrow-band signal that only intelligent life can make. These receivers routinely find narrow-band signals. How do we know whether or not these signals are extraterrestrial?

This question is one of the most difficult that face SETI researchers, and one of the most critical. Their equipment must reliably distinguish an alien signal from false alarms caused by bugs in the receiver, military radar, a telecommunications satellite, or a Caltech student prank. If researchers announced the greatest discovery of the millennium and then had to retract it, their entire enterprise would be discredited.

In the case of Project Phoenix, the SETI Institute's privately funded program to scrutinize nearby Sun-like star systems, an auxiliary telescope helps to sort out interference from alien broadcasts. Since Phoenix sifts through 28 million channels at a time, it unavoidably picks up transmissions from earthly activities. The auxiliary telescope, located at some distance from the main instrument, weeds out all but a small fraction of this confusing chatter. It does so by a precise comparison of the signal frequency received at the two antenna sites.

The handful of signals that make the cut are subjected to the tried-and-true litmus test used by every SETI observer since searching began in 1960: Move the telescope off the source and check whether the signal disappears. If it does, move the telescope back on source. The signal should reappear. If the suspected extraterrestrial transmission passes these tests, the telescope watches the target star until it sets, to see if the signal winks out on cue.

In 1995, Project Phoenix got underway at the Parkes radio telescope in New South Wales, Australia. Over four months, it examined over 200 star systems. While thousands of "signals" were found, none survived this confirmation obstacle course; all were terrestrial. But if one someday passed the tests, researchers would call on another, independent radio observatory to confirm the signal's extraterrestrial origin. It would take days or possibly weeks to convince the astronomers that they had indeed found an artificial emission from beyond the solar system. Only then would their discovery be made public. The world would be told.

Some people doubt that. They are skeptical about SETI's claims of openness. Influenced by persistent rumors that the federal government has kept alien UFOs and their bug-eyed pilots stacked up and hidden away, some people suspect that news of a detection would be squelched. This is unjustified paranoia. There is no policy of secrecy for any of the major SETI experiments, and if there were, it would be extremely difficult to enforce. Large numbers of researchers are involved, including the astronomers at independent observatories who would be called on to confirm a serious candidate signal.

The real problem is not that an alien transmission would, or could, be kept secret. Rather, scientists worry that a promising signal would generate so much excitement that it would be announced prematurely, only to prove later to be of earthly origin.

Meet the Press

Soon after that fateful, future detection, the researchers would call a press conference. Humankind would be faced with irrefutable evidence of cosmic company. Would there be panic in the streets? A groundswell of goodwill? Mammoth indifference?

When called upon to predict the reaction to a detection, sociologists and other prognosticating pundits turn to historical analogs. An obvious one is the 1938 radio adaptation of H.G. Wells's *War of the Worlds*. A Princeton University study suggested that more than a million listeners believed they were hearing actual news coverage of an invasion by hostile Martians. There was widespread panic and more than a few injuries.

Yet this fearful reaction had less to do with the fact of the extraterrestrials' existence than with their landing in our back yards. It's one thing to read in the morning papers that signals have been intercepted from a star system 100 light-years away, quite another to hear that aliens are exterminating the neighbors—especially at a time when world war is imminent.

Other analogies are also imperfect. Some historians hearken to publication of Nicolaus Copernicus's *De Revolutionibus,* when we were suddenly expelled from our central role in the solar system, or of Charles Darwin's *Origin of Species,* which implied that we were just another point on the continuum of life's evolution. Both shifted our perceived place in the scheme of things, and both generated outrage and uproar. Yet only a small, educated class read or understood Copernicus and Darwin. News of a SETI detection would quickly reach billions.

This would guarantee a wide range of reactions. A combination of media exploitation of outer-space themes and paranoia regarding UFOs has led the public to fully expect aliens on the airwaves. In a 1995 survey by Miguel Sabadell and Fernando Salamero of the University of Zaragoza in Spain, nearly four-fifths of the respondents said they thought that extraterrestrials exist. Gallup polls have shown similar results in the United States. Cosmic beings are part of the cultural milieu, the unconscious but shared mythology created by the media. There are bad aliens (*Alien*), good aliens (*E.T.*), and mechanistic aliens (*The Day the Earth Stood Still*). Public expectations and immediate reactions to a detection would be shaped accordingly. Each person would picture the beings at the transmitting end of a SETI signal as one of these Hollywood stereotypes.

Yet, despite the aggressive and violent behavior of many celluloid aliens, the surveys have suggested that the good ones make more of an impression. Four-fifths of the Zaragoza respondents said that extraterrestrial contact would be beneficial. (The majority of SETI researchers agree.)

For this reason, it can be assumed that most people would be excited and anxious to learn more about the extraterrestrials. Few would choose to abandon their daily routine in favor of rioting in the streets. But reactions would vary widely. Some would see the discovery as utopian ("Salvation is at hand!"), others as dystopian ("The end is nigh!"). Heated declarations would likely pit those with a favorable view of extraterrestrials against those who felt that their philosophical basis was under threat.

People beyond the media's reach are poorly informed about SETI and its premises. At best, they would remain indifferent to a detection. At worst, they might view it as a chance for the discoverers to monopolize a potentially important instrument of knowledge and control.

"I'll Call You Right Back"

That indifference might not last, particularly when thoughts turned to what, if any, reply should be sent. According to a protocol agreed upon by most SETI researchers, a reply should be transmitted only after international consultation. After all, if we are going to deliberately send a message to a distant society, it should represent the opinion of the entire planet. [Editor's note: A copy of the "Declaration of Principles Concerning Activities Following the Detection of Extraterrestrial Intelligence" is at www.seti-inst.edu/post-detection.html.]

Some think this is unrealistic. Freeman Dyson of the Institute for Advanced Study in Princeton, N.J. expects that anyone capable of cobbling together a transmitter and a satellite dish would start sending his or her own message to the aliens. In Dyson's view, Earth would respond to the extraterrestrials in a very human way: with cacophony. Donald Tarter of the University of Alabama has said that the researchers who detect a signal should immediately reply with the news of their success—and include a password to identify all future earthly communiqués from reputable sources. That way, if cacophony erupted, the aliens would at least know whom to tune in.

Concerns about a reply are probably moot. If the civilization is a thousand light-years distant, there is no urgency to reply. Furthermore, exactly what is said may not matter to an alien society whose level of technical and social sophistication dwarfs our own. And those who advocate silence for fear of hostile attack (as did Nobel laureate Martin Ryle two decades ago) ignore the fact that television and radar signals have already betrayed our presence.

Of course, when most people think about contact, they may be expecting something a little more personal than the detection of a narrow-band carrier wave from a sidereal source. Some people assume that detection would quickly be followed by arrival. After all, if Captain Kirk can transit the Galaxy in short order, the aliens presumably could as well. But the fact that we had heard them would have no bearing on whether they had heard us or were coming our way, a point that would undoubtedly be emphasized by researchers and officials during press conferences. But they may not be believed. Similar reassurances in the wake of the "War of the Worlds" broadcast were viewed as self-serving sops and an attempt to cover up the truth.

In reality, the alien transmission would probably seem pretty bland. Most searches, including those by the SETI Institute, are looking for continuous-wave or slowly pulsing signals that are easy to sort out from cosmic static, but can't carry much information. A variability in their frequency due to Doppler shift would reveal the axial and orbital rotations of the senders' planet (assuming they are on a planet). The aliens might also encode a little information in slow pulses, like a lethargic ham operator pecking out Morse code. The signals would tell us that the extraterrestrials were there, but not much more.

Once the scientific community had been alerted, however, it would do everything possible to investigate further. Optical telescopes would get busy trying to detect planets around the target star. Radio researchers would ask for money to develop the far more sensitive equipment needed to find any modulation, or message, associated with the detected carriers.

The Puzzle Palace

Suppose we were to receive modulated signals. Would we have any hope of understanding them? The answer would depend on whether we had picked up a signal intentionally beamed our way or had merely eavesdropped on internal traffic. In the latter case, the message might be impossible to unravel. Imagine Aristotle's puzzlement if he had been confronted with the task of decoding a 20th-century color-television signal.

If, on the other hand, the transmission were intended for reception by emerging technologies such as ours, the senders might ease our task by sending pictures or directions for deciphering their message. If we could crack the code of an alien signal, with or without their help, the impact on earthly society could be profound.

We are likely to find signals only if technological societies last for millions of years; if civilizations rise and fall quickly, they are unlikely to be present at exactly this moment in history. Consequently, any society we hear will be an old one, with a culture enormously beyond our own. It might provide us the opportunity to skip eons of history and leap into what otherwise would be a distant future.

Some of the researchers who have had the courage, or temerity, to consider the long-term consequences of a detection sense danger. A radio signal might not seem to pose the threat of an alien invasion à la *War of the Worlds,* but what if the aliens encouraged us to build innocuous-appearing machines that cloned these beings chemically or robotically? Could we become unwitting hosts to an alien sentience spreading at the speed of light?

This smacks of dark science fiction. More credible is the one-sided damage that could occur were we suddenly presented with information far beyond what we know. Would it be a disincentive for further inquiry of our own? Would we feel useless, without dignity in the face of superior knowledge? Understanding an extraterrestrial message could be like drinking from a fire hose.

According to a less pessimistic view, contact could help us to solve many of humanity's seemingly intractable problems. The fact that they exist, and therefore have survived the tumult of technological development, might give us the hope that we, too, can avoid self-destruction. Carl Sagan has predicted that contact would bring about "a new perspective on the differences we perceive among ourselves."

Like the public's initial reaction to a detection, the long-term consequences of a SETI success would be multifaceted. In the last four centuries, astronomers have forced humanity to accept that it occupies a humble, unremarkable place in a vast cosmos. A SETI discovery would remove the final cloak of superiority. We would know that we are neither culturally nor intellectually supreme, but simply one sentient society among many. We would no longer be special. Nor would we be alone.

Keeping the ETs Away

George C. Baldwin

George C. Baldwin insists that any attempt to contact extraterrestrial life could be dangerous to life on Earth. According to Baldwin, history has shown that whenever two different cultures meet, the result is conflict, subjugation, exploitation, and even genocide. There is no reason to believe that a first contact with an alien civilization would be any less hostile, Baldwin maintains. He asserts that the only possible good that can come of meeting aliens is the union of the entire human race against a common threat. Baldwin is a retired physicist and nuclear engineer in Santa Fe, New Mexico.

Confirmation that planets are indeed orbiting nearby stars, compelling evidence that water and volcanoes exist on some Jovian satellites, and most recently the news that simple forms of life may once have existed (still exist?) on Mars are truly revolutionary developments in astronomy. Meanwhile, marine biologists have discovered a third basic form of organism that thrives in the ocean depths under conditions utterly unlike those on land, deriving energy from carbon dioxide, sulfur, and hydrogen.

With these recent revelations at both cosmic and microscopic scales, "exobiology"—the marriage of astronomy and biology—is becoming an active science. It is not too soon to consider the implications. The rush of discovery should soften doubts that beings exist on other planets and that some may have evolved into advanced societies. I believe that, instead of a pleasant surprise, this latter prospect should be an occasion for serious concern.

Consider, for example, the boost all this activity on the exobiology front has already given to the search for extraterrestrial intelligence. When and if this search succeeds, idealists will grow eloquent and promoters of space travel will lobby for more money. I can hear them now.

"Imagine the benefits that will result if only we can establish communication," they will say, arguing that extraterrestrials (ETs) capable of communicating with us will, in all probability, not only be more advanced technologically but also have reached a utopian state that they will eagerly share with us.

"And, of course, with a civilization so advanced, the ETs may even have learned how to overcome the immense barriers of space and time that separate us!" they will enthuse. "How wonderful it will be to meet them face to face, to exchange ideas, and to learn from them!"

Cooler-headed physicists will counter that even a meaningful dialogue is impossible, since neither material objects nor messages can travel faster than light itself. The starry-eyed will reply, "Yes, space-time travel seems impossible today. But science fiction has a way of coming true. Whatever the human mind can imagine may eventually become possible."

I am willing to accept that interstellar travel could someday be realized—and that's precisely why I maintain that any effort to communicate with extraterrestrial life is fraught with grave danger! H.G. Wells's *War of the Worlds* may yet be proved prophetic. We should be worried not just with asteroid impacts and supernova explosions, nor just with the chance arrival of some "Andromeda strain" or incompatible gene borne by a Martian meteorite. What should arouse our concern is the very nature of life itself.

Let's reflect on the history of interactions among beings on Earth. Throughout terrestrial history, every contact of one civilization with another, human or otherwise, has ultimately resulted in the ascendancy of one and the subjugation, exploitation, and even extermination of the other. Even superficial differences among the human races—owing to skin color, religious beliefs, memories of long-ago conflicts—have triggered tragic clashes.

Given the amazing variety of life forms here on Earth, we can be certain (despite *Star Trek*) that extraterrestrial beings will be utterly unlike us in form—but not in their innate contempt for other beings. Just as on Earth, each life form will regard every other as inferior and, therefore, legitimate prey.

For example, consider our interaction with bees. They have a remarkable civilization—a model communistic society. Every individual in the highly organized hive knows its place and unselfishly does what it is capable of doing. Ages ago, while evolving the means for producing food, bees solved advanced problems in geometry, material science, engineering, and biochemistry. And what have we humans learned from contact between our two civilizations? With our "superior" technology, we steal the honey the bees work so hard to produce!

Fortunately, at present no other life form on Earth is capable of subjugating us—though bacteria and viruses are trying hard and may yet succeed.

"But," the idealists will argue, "being so much more advanced technologically, the ETs will have evolved into benign beings. Having mastered the struggle for existence, they will have no need of martial arts, nor will they need any of our resources. And with no history of past conflict with us, they will look upon us kindly and help us to solve our problems."

I grant these dreamers one point: any alien race able to contact us now is probably more advanced. But I do not agree that technical superiority guarantees benevolence. (Germany was the most technologically and scientifically advanced nation in the world when it launched World War I.) Nor do I concede that predation and exploitation are exclusively human traits. They are characteristic of all life, indelible genetic imprints that ensure some species will survive. Ants wage war with other ants, they exploit aphids, and we try to exterminate both.

We must also recognize the potential value to the ETs of Earth itself. It seems quite possible to me that a more highly advanced society, having exhausted its own sources of some rare element not yet depleted here, would want to commandeer our planet and lay claim to such resources.

The fact that we are light-years away from the nearest stars may give a false sense of security. For centuries Britain was isolated from the Continent by the English Channel, and the United States by two oceans, but new technologies overcame those barriers. With admirable foresight, some leaders realized that through those breached barriers would come predators bent on destroying an indigenous way of life. Over the objections of idealists, they prepared defenses and ultimately resisted the threat—but at great cost in lives and resources.

History and biology agree in telling us that if any good could come from confrontation with an extraterrestrial society, it would most likely be the uniting of discordant human societies to deal with a common threat. In the past we usually postponed preparation for conflict until confrontation was inevitable. It would be unwise to be unprepared for an extraterrestrial confrontation—but how would we divine the nature of our adversaries or their purposes and methods?

It is time to ask: Will the vast distances between stars always guarantee our safety as we continue to confront the many problems and tensions that exist here on Earth? When and if physical contact with extraterrestrial life becomes possible, is there any guarantee that we, rather than the ETs, will be the dominant life form—or will a few of us be consigned to a cosmic zoo while the rest are destroyed? To me, these unanswered concerns challenge the "wisdom" of advertising our presence by beaming radio, television, and telemetry into space, and by attaching "find us here" road maps to our interplanetary spacecraft.

How to Make a UFO Detector

Robert A. Goerman

According to Robert A. Goerman, it is believed that unidentified flying objects disrupt the Earth's electromagnetic field, causing electrical noise. If this is true, he notes, then many common household goods can be used to detect variations in electromagnetic fields. Goerman explains how television sets and transistor AM radios can be used to detect and monitor electrical noise associated with UFOs. Goerman has written many UFO-related articles for *Fate, Saga's UFO Report, Beyond Reality,* and *Official UFO.*

Static-plagued AM radio surrendered to FM's superior signal quality and stereo capability in the mid-1970s. Automotive passengers quickly learned to appreciate that "nonradio" audio equipment could be programmed to play only favorite music. Eight-tracks evolved to cassette tapes and to compact discs. Foil-festooned "rabbit ears" and rooftop television antennas seem like antediluvian relics. Black-and-white television imagery is today a quaint special effect. Television programs now arrive via coaxial cable conglomerates, by satellite dishes of every conceivable size and shape, by VCR, by DVD.

This electronics revolution faces many challenges. One is known as "electrical noise." Electrical noise is technically defined as a broadbanded electromagnetic field or interferential signal extending over a considerable frequency spectrum that is generated by various environmental and man-made sources. These include power lines and transformers, electric fences, lightning, corona and plasma, neon and fluorescent lighting, electric motors, high-frequency components, and electric arcs (a.k.a. sparks).

Also recognized as sources of electrical noise are Unidentified Flying Objects.

Saucer Static

The fact that electromagnetic forces and fields play an integral and quite inseparable role in genuine UFO encounters was realized early into the research. The National Investigations Committee on Aerial Phenomena (NICAP) established a subcommittee to evaluate these

Reprinted, with permission, from "Free UFO Detectors," by Robert A. Goerman, *Fate,* April 2000.

"electromagnetic effects associated with unidentified flying objects," which convened on July 30, 1959.

These "E-M Effects" fell into several categories, including automobile engine failure, radio and television interference, electrical burns (human, animal, ground), localized power failure, compass deviation, paralysis accompanied by a feeling of electrical shock, and corona discharge (e.g., sparks and lightning).

"The implications of these reports is that, whatever UFOs may be, they appear to affect electrical circuits under certain conditions," NICAP cautiously concluded in its 1960 findings.

To later researchers (most of them convinced that flying saucers were extraterrestrial visitors), those powerful magnetic fields and strange electrical forces represented accidental byproducts of an alien spacecraft propulsion system.

Beyond that, it was every expert for himself.

Richard Hall, in *The UFO Evidence* (1964), said: "Another conceivable explanation for the electromagnetic effects observed in the presence of UFOs is that some atomic device or weapon is used deliberately and selectively as a test or for other purposes."

Philip J. Klass, in *UFOs—Identified* (1968), said: "These electromagnetic effects are precisely the kinds that are to be expected if UFOs are plasmas. . . . A more common type of plasma effect is static or buzzing in a standard (AM) radio or disturbance of the picture on a TV receiver."

John A. Keel, in *UFOs: Operation Trojan Horse* (1970), said: "I now believe that the UFO phenomenon is primarily electromagnetic in origin."

Its existence acknowledged by skeptics and believers, this electromagnetic constancy sparked one question in the minds of pioneering ufologists: Could E-M "energies" betray the UFO location?

Inventive researchers got busy.

Automated Detection—The Magnetic Approach

Any mechanism sensitive to the approach of genuine UFOs (but unaffected by weather balloons, temperature inversions, or the planet Venus) could eliminate the challenge of personally scrutinizing the skies twenty-four hours a day, and could operate during inclement or uncomfortable weather without complaint.

Would a so-called "UFO Detector" really work?

This question gave birth to the "magnetic" school of thought. It was theorized that, given the reported spins and gyrations of magnetic compass needles during UFO sightings, as well as other clues, the earth's local and prevalent magnetic field was somehow affected when a UFO was present.

Devices were soon invented, or otherwise constructed, to detect these sometimes not-so-subtle changes in the prevailing magnetic field. Many individuals offered these devices (or plans to build them) for sale. The most extensive designs incorporated variations of what were termed "modified fluxgate magnetometers." Much simpler gadgets consisted of buzzers or bells that were connected by relay to a magnetic reed switch.

The crudest involved magnetic pendulums threaded through loops of wire, or simple compass needles rigged to work like a switch, and connected to a battery and buzzer. These latter contraptions were exceptionally delicate—they couldn't even be breathed on hard, let alone moved, without compromising their operation. It was recommended that these magnetic UFO detectors be nestled within the confines of glass jars, and kept safely maintained in a vibration-free environment.

One never could tell when a tree-top level UFO hovering directly overhead might just accidentally set one of these things off.

Or maybe not.

The Signal Is Noise

Try an experiment: fire up your computer and plug the phrase "UFO detector" into any number of Internet search engines to learn a disturbing truth. Almost nothing has changed in over three decades—until now.

Although not nearly as dramatic as paralyzed automobile engines and darkened headlights, in those days when rabbit ears magically snatched "The Invaders" out of thin air, and AM transistor radios surfed the airwaves, radio and television interference was *the* commonplace E-M effect associated with UFO encounters. Back when we had to get up to change the channel, or adjust the "vertical hold" and "fine tuning," these E-M effects were commonplace to the point of being taken for granted.

By definition, interference is *any* unwanted signal that precludes reception of the best possible signal from the source that you want to receive. It comes in three intensities: it may prevent or override reception altogether, it may cause only a temporary loss of signal, or it may affect only the quality of the signal.

Remember, these energies associated with the UFO phenomena have been rated strong enough to preclude powerful local radio and TV stations during the duration of the sighting.

What if—on purpose—the roles were reversed? What if we suddenly considered TV and radio broadcasts as nothing more than unwanted interference that competes with the UFO "signal"?

Nearly every household possesses at least one television set and several radios. Can these ordinary household appliances be fine-tuned to better detect and receive these mysterious electrical energies?

Welcome to the "electric" school of thought!

Television UFO Detection Technique

Important: your television must be connected to a VHF antenna—no cable or satellite dish.

1. Turn power on.
2. Adjust contrast to maximum.
3. Lower brightness 50 percent.
4. Select channel two, and you're all set.

As television is a visual medium, optimal results are obtained by maximizing the interactive clash between broad spectrum electromagnetic noise and an "occupied" channel two at 55.25 megacycles. With higher channels receiving transmissions above the 100-megacycle range, electrical noise becomes increasingly less effective.

Electrical noise will first appear as two or three white horizontal lines or intermittent bands, sometimes moving up to the top of the screen. The audio will also be affected. Stronger E-M noise will cover the television screen with flashing ripples or diagonal streaks of light. The strength of any interfering field drops rapidly with distance. At maximum levels, electrical noise will, in scientific terminology, drive the television "crazy." Many times the TV screen will glow stark white.

My original research began in the late 1960s with a vision of creating a "poor man's NORAD," involving a web of Television UFO Detection Systems in UFO flap areas.

Controlled experiments using terrestrial-based electrical noise were conducted on console TVs with large rooftop aerials. One successful preference employed a rotor-style motorized rooftop antenna whose direction could be conveniently changed and remotely-controlled by an indoor compass selector switch. This permitted the operator to track the interfering signal.

The Television UFO Detection System suffered some basic flaws. Although capable of monitoring electrical noise and UFO activity, these home-based consoles lacked mobility, and the effective detection range was totally dependent upon the size, sensitivity and elevation of the antenna. Every topographical obstacle to line-of-sight reception played a crucial role in limiting detection range. While the battery-operated black-and-white television sets of those bygone days offered some degree of portability, they were cumbersome, expensive to keep powered as they devoured carbon cells, and ignored all but the most powerful and local broadcast stations.

Something better was needed. Additional research supplied an answer.

The Virtues of AM

AM radio is the ideal instrument to detect and monitor electrical noise in any locale. It can be the ultimate UFO detector. Why? Because it meets four important criteria:

First of all: *it works!* UFO-inspired AM radio interference is a documented fact. In discussing automotive engine-failure UFO reports, NICAP's 1960 E-M study concluded that in "cases where the car was equipped with a radio playing at the time of the UFO appearance, the radio was the first to show any signs of disturbance." UFO-generated AM interference takes three forms: static, sounds, and those compelling "signals." Static is constant buzzing, like the sound of bacon frying. Sounds can include a roaring up and down the scale, speeded-up phonograph record noises, harsh shrieking, and high-pitched whistling. Signals are beeping, Morse-like "dots and dashes," or any interference structured into pulses.

Second: AM radios are simple to operate. Simply turn on the power, and select any quiet frequency between broadcast stations (preferably toward the high end at over 1600 KHz). You're all set. Remember: electrical noise *is* AM! It behaves exactly like an AM broadcast signal. Your low-fi AM receiver, unable to distinguish between a true broadcast signal and electrical noise, will pick up both, amplify both, and deliver both to your speaker. That is why we search only quiet frequencies between regular broadcast carriers.

The *FCC Interference Handbook* (on-line at www.fcc.gov/Bureaus/Compliance/WWW/tvibook.html), confirms the practicality of a simple AM receiver to detect weak electrical noise and interference within the home and neighborhood. FM, on the other hand, has a built-in rejection of noise, enjoying freedom from most electrical interference. By 1977, there were an estimated 205 million FM receivers, in 95 percent of all American homes—thus ending the AM radio domination.

Third: AM radios are portable. Lightweight and durable, pocketsize AM transistor radios are the rule rather than the exception. These little receivers perform flawlessly in the field and on-the-go, not at all impaired by movement or vibration.

Last but not least: AM radios are easily obtained. No soldering irons or complicated schematics to follow. No shipping and handling charges, or allowing 8–12 weeks for delivery.

Abundant and profuse—these words describe the great supply of inexpensive AM radios awaiting your experimentation. Good hunting!

CHAPTER 3

CLOSE
ENCOUNTERS

THE REALITY OF A FIRST ENCOUNTER WITH ALIEN LIFE

Paul Halpern

According to Paul Halpern, the public's fascination with extraterrestrials stems from the fact that most of the world has been thoroughly explored, leaving outer space as the only remaining frontier to discover. He notes, however, that humans tend to conceptualize extraterrestrials as being much like themselves, neglecting to take into account the likelihood of significant disparities in technology, culture, and languages. Assuming that the language barrier could be overcome, Halpern suggests that humans' greatest concern would be whether the aliens possess technological advancements that would enable them to solve all the world's problems. Halpern is a physics professor at the University of the Sciences in Philadelphia and the author of several books about the universe and space travel, including *The Quest for Alien Planets: Exploring Worlds Outside the Solar System,* from which this selection is excerpted.

Bug-eyed monsters with slimy, green skin and long, sharp talons, stomping muddy tracks through eerie marshes. Emotionless beings with raised eyebrows and pointed ears. Silver-coated androids running amok, blasting innocent Earthlings with powerful laser beams. Amorphous blobs that penetrate even the sturdiest of barriers. Doll-like, impish creatures with wrinkled skin, wanting only to "phone home." And telepathic energy fields that long for contact with corporeal races such as our own, silently hoping to become our friends.

Hollywood's Version of Aliens

These are some of the diverse images we have of extraterrestrials. They derive mainly from popular sources: speculative novels, plays, television programs, movies, and the like. For many years, a booming, Hollywood-based science fiction film industry has generated the costumes, props, and special effects needed to produce the illusion of life on other planets. Writers have spent their careers churning out page after page, describing imaginary beings from other worlds. Fictitious "alien" imagery is omnipresent—from *Star Wars* dolls in toy stores, to life-size cardboard cut-outs of *Star Trek* characters in book shops. And, years after its release, *E.T. the Extraterrestrial* is still one of the most financially successful and critically acclaimed children's films of all time.

Still, the public wants to learn more and more about mysterious "space creatures." A recently opened exhibit about aliens at the family vacation mecca, Walt Disney World, has already become one of the park's most popular attractions. Called the ExtraTERRORestrial Alien Encounter, it invites visitors to witness a life form from a distant planet. Thrill-seeking tourists sometimes line up for an hour or more for this strange, otherworldly experience.

The ride begins with the participants being whisked into a foreboding circular chamber where they are strapped into special seats and prepared for the experience of interstellar contact. There, they are instructed about a new scheme for the instantaneous conveyance of objects from planet to planet. A subsequent demonstration of interplanetary transport goes awry, and a hideous alien is brought to the chamber. Its horrid form suddenly appears within a glass containment vessel, right in front of startled onlookers. Soon—as any horror buff would expect—it escapes. As the audience gasps from sheer terror (or chuckles, as the case may be), it breaks out of its containment vessel. The room immediately becomes pitch black, adding to the suspense. State-of-the-art auditory, visual, and tactile effects enable frightened participants to feel they are being attacked by this hostile being. After the encounter is over audience members often leave the exhibit visibly shaken. (Children have been known to scream and beg their parents to be let out of the room well before the show is over.) Still, the crowds keep streaming into this popular attraction—located in the building that formerly housed another fictitious space show, the venerable Mission to Mars ride.

A Longing for Contact

Why does the public hunger to confront aliens? Why are media depictions of extraterrestrial life so captivating as shown by the popularity of such television shows and films as *Sightings, Alien Nation, A.L.F., The X-Files, Third Rock from the Sun, Aliens, Independence Day, Mars Attacks,* and several *Star Trek* series? Perhaps it is because—to borrow an expression from *Star Trek* that has grown into a cliché—space is the final frontier. We no longer send explorers halfway around the globe to seek out hidden civilizations and unknown peoples. Adventurers and anthropologists have raked the Earth over and over for signs of a yet undiscovered culture. Maybe there are still a few primitive tribes to be found, hidden from advanced society in remote jungle regions. In all probability, the cultural map of our world is essentially complete. The days of a Jonathan Swift writing about bizarre peoples (Lilliputians, etc.) living in unexplored parts of the globe are over. We need to set out into space to encounter beings and cultures that are radically different from what is already known.

In the centuries to come, as terrestrial culture becomes increasingly homogeneous, this longing for extraterrestrial contact will likely grow even stronger. With McDonald's restaurants in the Middle East and Coke machines throughout the Pacific Islands, trips to other planets (or imagined ones at least) may enable us to experience the exotic "lifestyles" radically dissimilar from our own.

When we finally do establish contact with alien life forms, assuming they exist, we will no doubt be amazed. Our current mental pictures of what extraterrestrials may be like stem mainly from depictions in fiction. Forming the stuff of cheap science fiction novels, television series schlock, and Hollywood B-grade movies, they are based purely on conjecture and longing. The reason for this lack of substance is simple. Authors' preconceptions are shaped necessarily by their own experiences. Even the most imaginative of writers cannot escape the fact that they were born on Earth and will likely never leave it.

Our utter lack of knowledge about alien life forms means if contact were to occur, we would be even less prepared than many of our ancestors were when they initially encountered societies more advanced than their own. In cases when less advanced peoples (of the Americas, Africa, and Oceania, for example) first came into contact with technologically superior nations (Europeans, for instance), they experienced strong culture shock. Often they were so confused and overwhelmed by their new circumstances they became ripe for exploitation—or extinction.

Imagine how the indigenous groups of remote Borneo, used to traveling exclusively on foot, felt when they saw their first airplane, or for natives of the Amazon basin, their first cameras. Almost immediately, their views of the world—of everything they knew and cherished—became disrupted beyond repair. It would have been virtually impossible for them to forget the experience and start over. Once their illusions were shattered, there was no turning back.

Considering the tremendous impact that the coming of Europeans had on aborigines in isolated parts of the globe, it is hard to fathom the far greater effects on both species that an interplanetary encounter would represent. This would especially be the case if the alien race possessed intelligence superior to ours. Should the other species be more advanced than we (and we need not assume this), then their existence would present a peril. We might only hope that the human race would survive such a strong challenge to its identity.

An Anthropocentric View

Am I suggesting that humankind would be "zapped" out of existence by alien weapons? Hardly. In my view, the far greater threat would be to our pride. We are used to being iconoclasts, to fending for ourselves in our own minuscule part of space. We think of ourselves as special—the lone bearers of the gift of conscious intelligent thought. Moreover, we traditionally consider ourselves the reason why the universe itself was created. The mere existence of extraterrestrials would shatter irrevocably this human-centered view of the cosmos.

Even in our space travel fantasies, we often imagine that our race will someday be anointed kings and queens of the universe, or at least of our galaxy. According to these geocentric visions, we would merely have to leave Earth, encounter aliens and flaunt our superior qualities to them, in a kind of cosmic debutantes' ball, and they would bow down to us.

Consider, for instance, the various *Star Trek* science fiction series. Though in these programs, many strange alien races are depicted, the bulk of those shown look and act very much like humans. Moreover, it is the prime goal of many of the fictional aliens presented to study, emulate, and even become humans. And it is most telling that humans are portrayed as dominating a future Galactic Federation, which is headquartered in San Francisco.

In truth, how likely is it that extraterrestrials would come to think of San Francisco, of all places, as their capital? Not very. It's a spectacular place, no doubt, with stunning vistas. But it's hardly central to places like Betelgeuse or Sirius. (It's not even central to Kansas City or Milwaukee, let alone the far reaches of the galaxy.)

In Isaac Asimov's well-respected *Foundation* series, on the other hand, the reigns of the galaxy are controlled by an empire centered on a distant planet called Trantor. Yet not even a visionary writer like Asimov could resist anthropomorphic temptations. The series gradually reveals that all of the galaxy's peoples can trace their origins back to a single planet—Earth!

Yet it's hard to fault science fiction for its pervasive geocentrism in making the assumption that Earth is the center of the universe. One might argue that one of the main functions of science fiction is to provide a thinly disguised version of earthly events. Often, it furnishes a useful forum for a "what if" look at our own history and culture. Thus we must consider its admonitions to be parables, for the most part, rather than prognostications.

The Language Barrier

If we were ever on the verge of meeting the real extraterrestrial denizens of our galaxy, we might not expect as easy an encounter as contemporary science fiction frequently suggests. Rather, we would have to brace ourselves for an experience for which there could be little true preparation.

If contact with aliens were to start through radio transmissions, our initial knowledge would be limited to what could be gleaned through broadcasts. Forming a picture of them in this manner would almost certainly prove challenging. Quite likely, in the beginning stages of contact, language barriers would make it difficult for us to obtain much information about them. Having no common ground to base our mutual communications would force us to struggle hard to fathom their messages, as well as to make ourselves understood.

All human languages, no matter how obscure, are grounded in certain commonalities. We each have a mouth, nose, two eyes, and two ears, as well as a mother and father. Each of us eats, drinks, and sleeps. Therefore, every language on Earth must have words or expressions for these body parts, relationships, and activities. When someone learns a new tongue, he or she might reasonably expect an answer to the question: "What is the word for this or that?"

Furthermore, human beings communicate with each other through standard, predictable mechanisms. In verbal and gesticulatory communication, we possess a fixed set of sounds and signs, from which each utterance must be fashioned. Moreover, a language mechanism in the brain shapes the types of grammar we can form. In written and transmitted language, we employ alphabets and/or pictographs that have similarity of origin.

Therefore, if anthropologists wish to translate a newly discovered language, they have the means at their disposal. They first look for similarities to known languages—in usage, grammar, and symbolic notation. Then, through these parallels, they form a "rosetta stone" connecting the unknown language with more familiar idioms.

With extraterrestrials, there could be no such assumptions of commonality. Although there would be some points of reference—the nature of the stars, the chemical elements, mathematics, etc.—there would be no shared experience to which we might allude. They could not draw from humanity's considerable lexicon, and we could not borrow from theirs.

Other Obstacles

Even if the language barrier were surmounted, we might find ourselves facing obstacles far more formidable. What if their entire way of thinking and being is beyond our comprehension and ours beyond theirs? Language is certainly no guarantee of true understanding. Even now on our own planet, where translators rapidly interpret the communications between nations, misunderstandings abound. And sometimes, sadly, these lead to war.

Yet for all the clashes between world nations, the situation between terrestrials and aliens would be far more volatile. Earth-dwellers, be they Africans, Europeans, Asians, or Americans, have many more commonalities than differences. As members of the same species, we harbor similar needs and desires. However, extraterrestrials, if they happened to be intelligent, would presumably possess radically dissimilar attributes. These severe differences could lead to profound difficulties, perhaps mutual hostility. Maybe, for example, the intonation patterns of human voices would horribly grate on certain alien races. Or the eating habits of some extraterrestrials might disgust and offend us, leading to attitudes, on our part, of extreme prejudice.

Let's be optimistic, however, for the purposes of this discussion. Assuming that these potential rifts could be somehow bridged—through patience and cooperation—our society would likely be amply rewarded by contact with an extraterrestrial civilization. In the ideal case, we would soon become privileged to share in the knowledge and wisdom of a vastly different race from our own. Through comparing ourselves with the other species, we would gain insight into what it means to be intelligent.

Yet even in the ideal case of benevolent, mutually enriching contact, there would still be grave dangers for the human race. Suppose the species we encountered happened to be vastly superior to us. We might be so humbled by their greater mental and physical capacities we'd lose our resolve. Their technology might be so superior to ours we might feel like children in comparison. Their understanding of the cosmos might be so comprehensive we would lose our sense of self-reliance and become dependent on them for all scientific information. In the extreme case, we might well decide that human scholarly endeavors would be pointless in the face of more comprehensive sources of knowledge. Our desire to explore the universe might well be quashed. In short, if the masquerade of human mastery of the universe were to end, there is danger that the resulting loss of pride could wholly rob our species of its sense of destiny and purpose.

Humanity's Biggest Hope and Greatest Fear

To return again briefly to the realm of fiction, consider Arthur C. Clarke's epic novel *Childhood's End*. In this dark tale, a group of alien visitors, called the Overlords, comes to Earth and cures all of its social woes. In a matter of decades, they eliminate the world's crime, warfare, and poverty, rebuild its cities, and institute world government. As they smother the human race with kindness and dazzle it with new technologies, it grows increasingly dependent on them. Ultimately, humankind becomes so spoiled by this treatment that it loses its special sparks of initiative and creativity. Science, art, and other forms of independent human expression all fall by the wayside.

In a poignant passage, Clarke describes how this intellectual decline takes place:

> There were plenty of technologists, but few original workers extending the frontiers of human knowledge. Curiosity remained, and the leisure to indulge in it, but the heart had been taken out of fundamental scientific research. It seemed

futile to spend a lifetime searching for secrets that the Over-lords had probably uncovered ages before. . . .

The end of strife and conflict of all kinds has also meant the virtual end of creative art. There were myriads of performers, amateur and professional, yet there had been no really out-standing new works of literature, music, painting, or sculpture for a generation. The world was still living on the glories of a past that could never return.

In summary, our biggest hope (however unrealistic) for an encounter with intelligent beings from another planet is that it would solve the outstanding problems facing the human race. On the other hand, our greatest fear for such an encounter is that it would solve the outstanding problems facing the human race. Although humankind would welcome connections with other cultures, it also thrives on challenges. One only hopes that once the shock of finding ourselves in some manner "inferior" to another species had passed, we would swallow our pride, regroup, and strive for new, loftier goals.

All this discussion is hypothetical, of course. Contact with extrater-restrials may be centuries, even millennia, away. And it is not necessar-ily the case that other forms of life in the cosmos possess greater or even comparable intelligence to ours. We must even consider the pos-sibility, however improbable, that there are no other intelligent life forms in space. In that case, our pride would be reprieved, but our lone-liness augmented considerably.

After all, it has only been in recent years that astronomers have firmly established there are planets around other stars. They have found at least a dozen new worlds, circling a number of distant suns. Though none of the newly found planets seem to present conditions favorable for life, these momentous discoveries, the focus of our explo-ration here, have increased our expectations that habitable worlds will soon be encountered. The possibility of finding living worlds has, in turn, bolstered our hopes—and apprehensions—about coming face to face with our extraterrestrial counterparts. Who are the aliens? Where are the inhabited worlds? We may find out soon enough.

WHAT TO DO IN A CLOSE ENCOUNTER WITH AN ALIEN

Richard F. Haines

In the following selection, Richard F. Haines gives suggestions as to what people should do if they suddenly encounter an alien or UFO. He urges people to remain calm, to carefully note the time and location of the encounter, to try to find another witness to the event, and to immediately record what happened by writing it down and describing it to a trusted person. In some cases, Haines writes, it may be prudent to try to escape from the aliens. Haines also discusses common experiences reported by abductees that may help other individuals to determine if they actually were abducted by aliens. Haines, a retired research scientist for NASA, has interviewed thousands of pilots and air traffic controllers about UFO sightings. He is the author of *Observing UFOs: An Investigative Handbook, Project Delta: A Study of Multiple UFOs,* and *CE-5: Close Encounters of the Fifth Kind.*

Any contact of the third or fourth kind is likely to be dangerous. This is, by definition, the unknown. But this does not mean you cannot prepare yourself physically and emotionally for the experience.

. . . Statistically speaking, it is a fact that unidentified flying objects (UFOs), or whatever they are, can appear at any place on earth and at any time of day or night. No place or time is safe from their presence, nor immune from their influences—and there are many different kinds of influences on humans both good and bad. UFOs do not respect any cultural boundaries, do not restrict their activities to people in particular language groups or to those who hold a certain pattern of beliefs. And they seem to display a kind of arrogant presence. What do I mean by this?

I have studied the subject of anomalous aerial phenomena for almost thirty years. Over this time I have noticed that these apparently metallic disks do not obey "established" laws of physics; in doing so

Excerpted from "Preparing for the Unknown," by Richard F. Haines, in *Making Contact,* edited by Bill Fawcett. Copyright © 1997 by Bill Fawcett & Associates. Reprinted by permission of HarperCollins Publishers, Inc.

they continue to confound scientists. Some of these vapor-shrouded objects can produce strong radar returns. Others are seen by multiple eyewitnesses but remain invisible to radar. They often fly "too close" to airplanes; indeed jet interceptor pilots seem to be taunted by them. Official U.S. government records have acknowledged that very bright unidentified balls of light incapacitated several electrical systems on board an American-built F-4 Phantom jet of the Imperial Iranian Air Force in September 1976 sent up to chase them. The lights clearly outperformed the jet aircraft. After their aerial game of cat and mouse was over, the lights accelerated out of sight to the west.

Other UFOs may hang silently in the night sky near a farm house blinking what look like anticollision beacons until someone comes outside with a shotgun. Seven cases have been reported where someone heard their bullet ricochet off the surface of the object, like "metal striking metal." The object sometimes emits a small object or a loud sound or a ray of light before quickly shooting upward out of sight. Still others progress majestically across the daytime sky in formations of ten, twenty, or more at a time. It is as if they are toying with us, at least acting as if they are invincible.

Based upon these kinds of typical UFO responses reported from around the world, we should ask, What should you do if you are suddenly confronted by a UFO? This is the subject discussed here. I will suggest some practical things you should do, each of which is related to different ways UFOs manifest themselves. These UFO manifestations fall into several groups; one's response should vary depending upon which kind of event occurs. These groupings are I. distant sightings, II. close encounters, and III. direct bodily involvement.

I. Distant Sightings

An example of this kind of UFO experience would include a nighttime sighting of an anomalous light passing in zigzag fashion across the sky, an obviously distant grouping of multicolored objects cavorting like a military acrobatic team, a radar contact in an air traffic control center, or a distant cloud that seems to coalesce into a smooth metallic object that then flies away. There are scores of other possibilities as well. In the suggestions that follow it is assumed that the person can see, hear, and/or feel the presence of the UFO but is not in any clear or present danger from it.

What to Do: First, do not panic. There is no reason to panic because it will do no good and only may increase the risk of personal injury. Just try to stay calm and in the same place throughout the entire sighting. *Second,* if you think you are perceiving something highly unusual in the sky (or under water), mark your present location in some way. Also take careful note of the direction you are looking in. Use all available visual landmarks and a compass if possible. This will greatly assist others later in studying the phenomenon. *Third,* if possible, try to find someone else to come see the phenomenon with you, but if you are successful, do not compare notes with him. Allow him to record his own experiences. Most UFO sightings last about five minutes or less, so this step may not be possible. *Fourth,* get a pencil and paper and write down exactly what happened. Record the time, date, and place, and make sketches. Try not to interpret what you experienced; leave that for others later. Write down only objective facts.

II. Close Encounters

The author of this now-familiar term, *close encounter,* Dr. J. Allen Hynek, was a professional astronomer and professor of astronomy at Northwestern University near Chicago. He was hired by the U.S. Air Force relatively early in his career to provide scientific analysis services to their Project Blue Book, a collection and analysis of anomalous atmospheric phenomena, which many military personnel had been reporting since World War II. He once told me that when he first began this consulting work, he didn't believe UFOs existed. "This job probably couldn't amount to much," he once said. Little did he realize that he would be given access to such startling classified evidence positively supporting the existence of UFO phenomena that he would have to change his mind about them. Although initially skeptical that such objects could exist, or at least be visiting earth across the vast expanse of space that surrounds us, eventually Dr. Hynek did change his public position. When Project Blue Book was terminated (December 17, 1969), Hynek lost this consulting work and went on to found the private Center for UFO Studies near Chicago. He also developed the first useful taxonomy of sightings, which continues to be used today—a classification scheme that includes the idea of the "close encounter."

Dr. Hynek found it useful to group close (UFO) encounters into three separate categories, which are defined below. As before, I will provide suggestions on how to respond if you should find yourself in each kind of close encounter; my comments are based on working with many people over the years who claim to have had such an encounter as well as what happened to them afterward. Of course one cannot know in advance, or even during the experience, exactly what kind of close encounter one is having. . . . But if you should find yourself in this situation and have a general understanding of these categories, it may help you to decide what is happening and what you can do about it.

CE-1: Close Encounter of the First Kind

According to Hynek's definition, the first kind of encounter is one in which the unidentified flying object (UFO) is seen at close range but produces no permanent interaction with the environment (other than some possible psychological trauma to the observer). He defined "close" as being within about several hundred feet of the object, craft, or phenomenon. . . .

What to Do: First, do not panic! Such self-defeating behavior will only increase the risk of personal injury and lessen your ability to remember details accurately. *Second,* stay put. Don't run or try to hide except to protect yourself from debris or other objects that may be flying through the air. *Third,* try to concentrate mentally on as many details as possible. If you are having a "typical" CE-1, your senses may be literally filled to overflowing. Your mind may seem to be in a confused state, your adrenaline pumping. And you may feel like immediately running and hiding. But don't do it. It won't make any difference anyway to the phenomenon—intelligent or unintelligent, animate or inanimate. If what you are seeing is a craft under the guidance of highly intelligent beings and it is within a couple of hundred feet, "they" know of your presence and probably will carry out their objective anyway. And if "they" are only a natural phenomenon, such as a luminous plasma that is "riding" on a nearby electromagnetic field line, it probably won't matter very much whether you are close or far away. *Fourth,* after the encounter is over, mark the spot with a pile of stones, broken tree branches placed in a cross, a handkerchief tied to a twig, and so on. *Fifth,* go home and write down everything you can remember, including all seemingly unrelated events; make drawings. For example, if you come home later than you expected, note the times involved (when you left and when you returned), or if you suddenly notice some new scar or other skin blemish, note it in your journal. *Sixth,* try to find someone you trust and care for and tell him or her exactly what happened to you. It's okay to insist on confidentiality from her at this point. Such sharing will help you get the experience outside yourself and make it more objective. *Seventh,* telephone or write to one of these organizations to report your experience:

Mutual UFO Network
103 Oldtowne Road
Seguin, TX 78155

J. Allen Hynek Center for UFO Studies
2457 West Peterson Avenue
Chicago, IL 60659

They will treat you with respect and keep your information confidential if you so indicate. *Finally,* if you think you can tolerate the social ridicule and curiosity that usually accompany reporting a UFO sighting, you may want to contact the authorities. Call your local police, newspaper, radio station, sheriff's department, or the Federal Aviation Administration (FAA) to find out if others reported seeing the same thing. But then be prepared to be ridiculed. In spite of the fact that about one-half of America's adults now believe in the existence of UFOs, our society is not yet ready to respond with the compassion or acceptance that is needed toward those who have actually encountered one.

CE-2: Close Encounter of the Second Kind

A Close Encounter of the Second Kind refers to events similar to the first kind except that definite physical effects are noted on both animate and inanimate material. For example, vegetation is scorched or pressed down; tree branches or grass stems are bent or fractured; various-shaped impressions are made in the ground; soil may be affected in other ways (e.g., dehydrated, scorched, irradiated, covered with some residue); animals may act with fright. Inanimate objects such as automobiles and airplanes may be interfered with in different ways. For example, radios may suddenly change stations or experience static interference, car engines can stop altogether, compasses may rotate, and direction finders cease to operate. A broad spectrum of so-called electromagnetic effects have been reported.

Your response to a CE-2 should be little different from your response to a CE-1 or a distant sighting. Referring to the previous section, follow all eight of the steps listed but add a *ninth*, which will help protect you from possible injury. If you should find some unknown artifact(s), materials, substances, or ground impressions, do not touch or disturb them in any way. Not only will you protect yourself from possible injury but you will help safeguard the evidence for others. If it is raining, try to locate a large plastic tarpaulin (or painting sheet) to cover the spot where the evidence is.

CE-3: Close Encounter of the Third Kind

A Close Encounter of the Third Kind involves seeing a UFO at near range in which one or more humanlike creatures are seen or associated with the object in some way. Interestingly, while the literature contains many such stories, the human witness is seldom affected in any significant way. More commonly the creatures go about their business as if no one is present, even ignoring the humans altogether.

What to Do: First, do not panic, but instead try to remain calm and as observant as possible. *Second,* if you are approached by "them" and can walk or run away, do so. Don't wait. If at all possible, don't permit yourself to be captured by the beings. Unfortunately this may be less up to you and your response and more up to "them," as will be mentioned in the following section. Most likely they will leave you alone. *Third,* try to remember as many details as you can. You will probably be very excited or confused; try to use a mental-association technique to help you recall details later. Associate a visual detail with a letter of the alphabet, such as the following:

A = *a*liens (creatures seen, about five total, one taller)
B = *b*ees (they moved around in an irregular manner)
C = *c*ounterclockwise rotation, disk, flew away vertically
D = *d*ipped down as it approached
H = *h*umming sound

L = *l*ights, many colors, red and blue predominated
M = *m*etallic silver suits, worn by creatures
and so on

III. Direct Bodily Involvement

A CE-4 [Close Encounter of the Fourth Kind] is still poorly understood and is shrouded in fear and skepticism in most Western societies. In a typical encounter in which someone claims to have been abducted, we will find a broad array of symptomatology: unexplained body marks, abnormal emotional reactions, an unusual altered state of consciousness, lucid dreams, out-of-body vision(s), obsessive-compulsive behavior, personality changes, reception of previously unknown information, and other responses. It is partially this breadth of symptoms that has confused investigators and inhibited most physical scientists and physicians from getting involved.

So how can you know whether you have been abducted? This is not a simple question, and its answer must be found in a comparison of the experiences of many others who claim it has happened to them. Importantly, a large majority of these people describe much the same set of events. It is this particular diagnostic "sign" of event-consistency that can be used to determine the likelihood that an abduction has taken place. Thomas E. Bullard has said, "A depth of integration holds the themes of the abduction story together in a plausible whole. The experience lasts a lifetime, and shifts focus in keeping with changes in the life cycle, especially the reproductive cycle." In other words the high degree of consistency found between the details of the claimed abduction scenario and one's lifestyle changes are striking and suggest that an actual, physical event has taken place.

Abduction Episodes

Thus, if you have experienced at least seven or eight of the following nine "episodes," as I call them, there is reasonable evidence to believe that you have had a close encounter of the fourth kind, sometimes called an alien abduction. These episodes are distilled from reading the literature and working with many people who claim to have had a CE-4. Other researchers offer other signs as well.

What are these general episodes? They include the following:

1. An *alerting-orienting stimulus,* which diverts your attention from some current, ongoing task to a new direction or event. Usually the stimulus consists of an unusual light, noise, vibration, or electrical-shock sensation. There is reason to believe that if one does not permit one's attention to remain diverted in the new direction, the other (abduction) events will not take place.

2. *Capture;* i.e., those events surrounding one's transport or abduction from the original environment to a new, unfamiliar environment,

often a craft or relatively small enclosure. A currently debated issue is whether or not transport is physical or only mental. There is fascinating evidence on both sides.

3. *Communication* and *special messages* with/from the "beings." Any mental (telepathic), auditory, or written transfer of information to you qualifies here. Interestingly, there are documented cases where the person develops new talents, understandings, and "capabilities" suddenly following the claimed abduction.

4. A *tour* of the new place you find yourself in. Some abductees interpret this episode as being given to them in exchange for their cooperation. They are often shown highly advanced technology, architecture, displays and controls of some sort, as well as a host of other unfamiliar things.

5. A *personal examination* of some type; it could involve scrutiny of one's body, mind, or psyche. It is usually performed ruthlessly in comparison with terrestrial medical practice. The "visitors" do not seem to realize that human beings feel pain, nor do "they" appear to respond to our humor.

6. *Travel to and from a second, larger craft or enclosure;* this episode is interesting for several reasons. First, it is not a theme found in very many science fiction books or movies (and therefore probably does not originate in them), and second, very often this episode has a different thematic character from the abductee's preexisting lifestyle, interests, or belief system. The place where one is allegedly taken is sometimes described as mythological in character, with Greek-like statues and columns, giant animals, and vaporous beings. The larger environment itself is always completely enclosed and may appear less technological (e.g., a gardenlike valley with an artificial sky/roof) compared with the original craft or enclosure/cave.

7. A *tour of a second environment* is given. Here one may be escorted to different locations, during which time the beings may communicate telepathically. As before, the tour episode seems to be offered as an inducement to cooperate as much as to educate.

8. The *return* episode includes all events surrounding how you were brought back to your starting point. My work has shown that one is returned in the identical mode of transport that one was originally taken to the second (large) environment, almost like running a movie backward.

9. The *aftermath* episode includes all of the physical, psychological, physiological, psychic, and spiritual life changes that are experienced after the abduction. Of course some of these sequelae can serve as medical and scientific evidence for the occurrence of a traumatic event(s) and as such are very important to have been examined. As many hundreds, if not thousands, of people have discovered, their personal aftermath events can last a lifetime.

What to Do: It is difficult to provide a list of specific procedures to follow for several reasons. For one thing, each person's abduction experience is somewhat different, and for another, it is most likely that you will not remember very much of your encounter until much later. Some people begin to recall fragmented bits of details only after many years. It is common for some type of trigger event to occur to cause a memory to jump into consciousness. You are suddenly confronted with a bizarre recollection that seems to be based on a real event but that makes no sense by itself. You may simply experience a slowly growing awareness of some of the earlier abduction episode(s) at a conscious level. Lucid dreams and/or precognitive visions of future events may also occur. Because of the diversity of possible responses it is difficult to advise you exactly on the best course of action.

Two Courses of Action

It is fair to say that you have two general courses of action in seeking help with your experiences. In one, which I call the *clinical course,* you would meet with a licensed psychologist or psychiatrist in order to work out the best stress-coping techniques for you. Your primary objective would focus on establishing your psychological health and well-being. In the second course of action, which I call the *research course,* you would work closely with a trained UFO investigator who understands what you have gone through and who also understands something about the rich diversity of UFO phenomena. Your mutual objective would be oriented more toward understanding the core identity of the phenomenon and its impact on yourself and mankind. To help keep from biasing you, the investigator would probably not tell you many details at first but would only gradually disclose relevant facts as your meetings continued. But whichever course you should follow, be prepared for a long and intense interaction, where your deepest beliefs and attitudes may be probed and challenged, where your personality may be tested and analyzed. Be prepared to become a new person.

Therefore, if you suspect that you may have been abducted, *first,* keep a personal diary as you begin to recall earlier experiences no matter how bizarre they may seem. Such records can be of great value to investigators, who can help you work through your traumatic aftermath. But don't expect everything to suddenly become clear. It may require months or years of living with turmoil inside yourself. Sometimes you may even come to think that you are losing your mind. But persevere in what you know to be a psychologically real and intensely personal experience.

Second, tell someone you trust and care for about your suspicions. Tell him as much as you can and ask for his support. Do not keep your experience to yourself.

Third, if you believe that the abduction occurred recently and you notice some bodily scar or other symptoms, for instance, persistent double vision, metallic taste, ringing sounds, disorientation, unsettling dreams, blood discharge, persistent or abnormal thirst, an oversensitivity to light, skin burns, abnormal cardiac responses, and so forth, you should seek assistance from a medical professional regardless of whether he or she believes in the reality of the C-4 or not. Unfortunately it is true that most physicians in America will jump to a psychiatric explanation for most of your claimed aftermath conditions. You should decide in advance how much you are willing to tell your doctor about your encounter. Fortunately there are some medical practitioners supportive of such claims who are also experienced in working with people who think that they have been abducted. Some practitioners are trained in the use of special methods and techniques to unlock your memory. . . . You should recognize that use of hypnotic regression may or may not actually help in recovering these memories, depending upon many factors. If you think that the abduction took place a long time ago, it may not be as important to obtain an immediate medical examination or counseling. Remember that it is your physical well-being that needs to be safeguarded before documenting the possible reality of the event.

Fourth, don't try to deny the reality of your memories, no matter how strange and impossible they may seem to be. Denial will only embed them deeper into your subconscious. And if, somehow, you should discover that your abduction was only an imagined event with no possible basis in fact, your subconscious will be able to find suitable ways of catharting the associated emotional baggage you may have been carrying around. In other words, time will have a way of healing you of your scars.

Personal Observations

After working with many people from all walks of life and from many different countries of the world who claim to have seen an anomalous aerial phenomenon or of being taken against their will to a very strange-appearing place, and after studying the reliable literature, I am convinced that they are describing a highly consistent phenomenon that has been with mankind for a very long time. I have little doubt that to be forewarned and educated about the things that can happen to you is a good thing, a useful thing. I hope that you may never need to apply these suggestions, but if you do, perhaps you will be better off for having considered them.

THE TRUTH ABOUT UFO SIGHTINGS

Philip J. Klass

Leading UFO investigator Philip J. Klass is the author of several books about unidentified flying objects, such as *UFOs: The Public Deceived, UFO Abductions: A Dangerous Game, Bringing UFOs Down to Earth,* and *The REAL Roswell Crashed Saucer Coverup.* Klass asserts in the following essay that in five decades of UFO investigations, no credible evidence has been produced to support the theory that UFOs and extraterrestrials have visited Earth. In fact, he maintains, the only evidence supporting extraterrestrial visitation are "unexplained" UFO sightings, which if investigated thoroughly enough, can be explained by some ordinary event. Therefore, UFO investigators have little incentive to examine "unexplained" cases too closely, he contends.

Despite widespread media coverage of UFOs (Unidentified Flying Objects), one important and well-demonstrated fact is seldom mentioned: At least 90 percent of all UFOs are really "Misidentified Prosaic Objects," or MPOs. More than one-third of all UFO reports are generated by bright planets, stars, meteor-fireballs, and even, occasionally, by the moon.

For example, the March 20, 1975, edition of the Yakima, Washington *Herald-Republic* carried a front page feature stating that three credible local citizens reported seeing a bright UFO in the western sky around 9 p.m. and watched it for about 45 minutes "until it disappeared." One man, who described the UFO as being cone-shaped with a "greenish-bluish light at the top and a sort of pale flame light at the bottom," said he had never seen anything like it before. Not surprisingly, the front-page story prompted other Yakima citizens to look for a UFO that night.

The next day's edition reported that many more persons had called in to report seeing the UFO, which had returned to the western sky that night at about the same time. The following day, the newspaper reported that its staff had been "swamped by calls by people mystified and convinced they were seeing an alien craft from outer space."

Fortuitously, one of the callers was an amateur astronomer who, after reading the two previous newspaper articles, had decided to take

Reprinted, with permission, from "A Field Guide to UFOs," by Philip J. Klass, *Astronomy*, September 1997.

a look at the UFO. He reported it was the then-bright planet Venus. Commendably, the *Herald-Republic* reported the UFO's identification on its front page, whereas many newspapers would have buried the explanation at the bottom of page 18.

Little Credible Evidence

There is scant scientifically credible data on the relationship between MPOs and UFOs. Since 1969, when the U.S. Air Force formally ended its 20-plus-year investigation of UFO reports and closed down its Project Blue Book office, nearly all of the people who now investigate UFO reports want to believe that some UFOs represent visitations by extraterrestrial spacecraft.

Despite many reports of UFO landings and persons who claim to have been abducted and taken aboard a flying saucer, no one has yet come up with a single credible physical artifact to confirm the ET hypothesis. The "physical evidence" found at alleged UFO landing sites typically consists of broken tree branches, or small holes that could be the work of wild animals, or a hoaxer. Not one of the many so-called "abductees" has come back with an ET souvenir or any new ET scientific information which could be verified to confirm their tale.

Because nearly half a century of UFO investigation has failed to yield a single, scientifically credible physical artifact, the only evidence supporting the ET hypothesis rests entirely on "unexplained" UFO cases. If an investigator is unable to find a prosaic explanation for a UFO report, believers in the ET hypothesis cite this as evidence that the UFO report was generated by an ET craft.

Because nearly all of the persons who now investigate UFO reports are eager to find "unexplainable" cases, this gives them scant incentive to conduct a rigorous investigation of a tough case. During my more than 30 years of investigating UFO reports, several cases required many months of part-time effort to find a prosaic explanation and one required more than two years.

CUFOS

A notable exception is the UFO investigative work of Allan Hendry, who became the chief investigator for the Center for UFO Studies (CUFOS), shortly after he graduated from the University of Michigan in 1972 with a B.A. in astronomy and illustration. CUFOS had just been created by J. Allen Hynek, who then headed Northwestern University's astronomy department.

In the late 1940s, the Air Force hired Hynek as a consultant after it discovered that bright celestial bodies generated many UFO reports and that even experienced military pilots sometimes chased after a UFO that turned out to be a bright planet or star. At the time, Hynek was teaching astronomy at Ohio State University—not far from the

Project Blue Book offices in Dayton. In 1969, when the Air Force decided to close down Project Blue Book, it terminated Hynek's contract. Several years later he created CUFOS and hired Hendry as its full-time UFO investigator.

For Hendry, with a long-standing interest in UFOs, it was an exciting opportunity. Although Hynek had been a hard-nosed skeptic about the ET hypothesis when he was first hired by Project Blue Book, in his later years he became more "open-minded." He hoped that CUFOS could conduct a competent scientific investigation into the UFO mystery. To encourage UFO reports by law enforcement officers, who were considered to be more reliable than the general public, CUFOS obtained a toll-free 800 telephone line whose number was given to many police departments.

During the next 15 months, Hendry personally investigated 1,307 UFO reports submitted to CUFOS—many more cases than any other investigator up to that time, or since. The results were published in 1979 by Doubleday/Dolphin in a book entitled *The UFO Handbook: A Guide to Investigating, Evaluating and Reporting UFO Sightings.*

Hendry's book—regrettably long out of print—provides the most recent scientifically credible data on the many different "trigger mechanisms" which generate UFO reports. Earlier Project Blue Book data is criticized by some who charge that the Air Force's eagerness to explain away UFO reports resulted in unrealistic explanations. Hendry's data is not vulnerable to such charges because he admitted he would like to find evidence that some UFOs were ET craft.

Of the 1,307 UFO cases that Hendry investigated, he found prosaic explanations for 91.4 percent of them, leaving 113, or 8.6 percent, unexplained. However, Hendry conceded that 93 of these 113 unexplained reports had possible prosaic explanations. This left only 20 cases, or 1.5 percent of the total, as seemingly unexplainable in prosaic terms. Is that evidence that some or all of these 20 unexplained cases involved ET spacecraft?

Hendry wisely resisted making any such claim. In the closing pages of his book, he admitted that rigorous investigation alone is not always sufficient to find a prosaic explanation, which sometimes depends on "sheer luck." (Based on my own investigations over the past 30 years, I fully agree.) Another factor, which Hendry does not mention, is that with his other CUFOS duties, which included producing a monthly publication for subscribers, he could spend an average of only two hours in his investigation of each case. Some of the cases I have investigated have required many dozens or hundreds of hours of effort to find a prosaic explanation.

Early in my own career as a UFO investigator, I was "taken in" by several hoaxers who seemed at first to be honest. I suspect that at least several of Hendry's "unexplainable" cases are hoaxes and that he was too trusting, as I had been.

In the concluding pages of Hendry's book, he commented: "How can I be sure if my remaining 'UFOs' aren't simply Identified Flying Objects [i.e. Misidentified Prosaic Objects] misperceived (sincerely) to the point of fantasy? The emotional climate about the subject (as revealed by Identified Flying Objects) appears to be adequate to support such a hypothesis for a great many UFO situations, if not all. . . . With our current inability to fully draw the distinction between real UFOs and IFOs, fantasies or hoaxes, coupled with a heated emotional atmosphere, I can only assert that it is my feeling that some UFO reports represent truly remarkable events." But Hendry admitted that "while science may be initiated by feelings, it cannot be based on them."

Celestial Bodies

Hendry's investigation showed that nearly 28 percent of all UFO sightings reported to CUFOS proved to be bright stars and planets. Hendry noted that UFOs that turned out to be celestial bodies were often reported to "dart up and down," to "execute loops and figure eights." Occasionally the celestial UFO was reported to "meander in square patterns" or "zigzag." In 49 cases triggered by a bright celestial body, the witnesses estimated the UFO's distance at figures ranging from 200 feet to 125 miles.

When I first entered the UFO field, I would have challenged the idea that an intelligent person could mistake a bright celestial body for a UFO that was "following them." But numerous incidents, some involving law enforcement officers, have convinced me otherwise. If a person driving in a car sees a bright celestial object ahead and suspects that it might be a UFO, and accelerates to try to get closer, no matter how fast the driver goes, he/she cannot seem to gain on the UFO. If the driver then stops and gets out of the car, the "UFO" seems to halt also, because it is getting no bigger or smaller. Now, if the driver decides to return home, the "UFO" seems to be following the car because it remains the same size and brightness.

It might seem surprising that 22 of the UFO sightings reported to CUFOS turned out to be triggered by the moon. Early in my career as a UFO investigator, I was challenged to explain an incident that had occurred with a Navy aircraft crew on February 10, 1951, while en route from Iceland to Newfoundland. The Project Blue Book files listed the case as "unexplained." After careful study of the crew's report, many hours of investigation, and a bit of luck, this UFO could be identified as the upper tip of a crescent moon which was barely visible at the horizon.

Hendry's investigation showed that nearly 18 percent of all UFO reports were generated by advertising airplanes, which carry strings of lights that spell out an advertising message. When seen at an oblique angle, their strings of flashing lights are perceived as being saucer-shaped. When the pilot decides to turn off the lamps and go home, observers report that the "UFO mysteriously disappeared."

Meteors and Rockets

On June 5, 1969, the flight crews of two east-bound airliners flying above 30,000 feet near St. Louis, Missouri, as well as a military jet fighter pilot, reported a near mid-air collision with a "squadron" of hydroplane-shaped UFOs. The incident occurred around 6 P.M., in broad daylight. The military pilot reported that at the last minute the UFOs seemed to maneuver to avoid a collision—which could be interpreted as evidence that these objects were "under intelligent control."

Thanks to Allan Harkrader, an alert newspaper photographer in Peoria, Illinois, who was able to take a photo of the UFOs, they could be easily identified as flaming fragments of a meteor-fireball blazing through the atmosphere on a horizontal trajectory. As a meteor enters at very high speed, it ionizes (electrifies) the surrounding air, creating a long, luminous tail. The two airline flight crews and the military pilot, understandably, assumed that the luminosity was the result of sunlight reflecting off metallic objects.

Based on Harkrader's photo and other reports from ground observers, scientists at the Smithsonian Astrophysical Observatory were able to calculate the approximate trajectory of the fireball. They concluded that the fireball actually passed about 125 miles north of St. Louis, at an altitude many thousands of feet higher than the two airliners and the jet fighter. The same fireball fragments were seen by a private pilot who had just landed in Cedar Rapids, Iowa. He filed a report on the incident with the Federal Aviation Administration's office in which he estimated the objects had flown directly over the airport's east-west runway at an altitude of only 1,000 feet. His distance estimate was in error by roughly 100 miles.

On March 3, 1968, at approximately 8:45 P.M., a group of three people in Nashville, Tennessee, saw what appeared to be a giant saucer-shaped UFO pass over them in eerie silence at an altitude estimated to be only 1,000 feet. In a detailed report later submitted to the Air Force, they reported seeing many rectangular windows, illuminated from inside the craft.

The Air Force also received a report of the same incident from a group of six people in Shoals, Indiana, 200 miles to the north. They described the UFO as being more cigar-shaped, with a rocket-like flame in the rear, but they too reported seeing rectangular windows, illuminated from inside. They reported its altitude was "tree-top level."

This UFO proved to be the flaming reentry of a Soviet rocket that had been used to launch the Zond-4 spacecraft on a simulated lunar mission. As the rocket reentered at high speed, it broke into fragments which were heated to incandescence, which is what the observers in Nashville and Shoals saw. Once they concluded that this was a UFO, then the flaming fragments "became" windows illuminated from inside the craft. And their brains, unwittingly, supplied details of the craft's shape—based on what the observers had earlier read or heard about the shape of UFOs.

One of the universal characteristics of UFO reports generated by fireballs or reentering space debris is that the object always seems to be much closer than it actually is. In Hendry's book, he reported that nearly 9 percent of the 1,307 UFOs reported to CUFOS proved to be fireballs or reentering space debris.

Other Objects

Hendry, who retired from "UFOlogy" shortly after his book's publication and has never returned, discovered there are many, many trigger mechanisms for UFO reports besides bright planets, fireballs, reentering space debris, and the moon. These included hoax hot air balloons, weather and scientific balloons, missile launches, birds, and kites, to cite but a few.

Several years ago, shortly before I was to give a UFO lecture to the Seattle chapter of the Institute of Electrical and Electronics Engineers, I was standing outside the lecture hall chatting with several attendees. Suddenly, one of them pointed skyward and said, "What's that?" I looked up and there was a small, orange object which seemed to be hovering at an altitude of several thousand feet.

Someone said: "It looks like a kite." I responded: "No, it's much too high to be a kite, maybe it's a weather balloon reflecting the nearly setting sun." The other party responded: "It can't be a balloon, it's not moving." Suddenly, a third man spoke up: "I think I have some binoculars in my car," and he hurried to get them. He returned with the binoculars, took a brief look and said: "It's a kite." When I viewed the UFO through the binoculars, I agreed.

Were it not for the happenstance that this man had a set of binoculars, I would have to admit that I saw a "UFO" in Seattle that appeared to be too high to be a kite and too stationary to be a weather balloon. But it was not doing anything extraordinary—or extraterrestrial.

For 45 years I have been writing for *Aviation Week & Space Technology* magazine—for 34 years as one of its senior editors and since my "active-retirement" in 1986, as a contributing editor. The magazine has published more articles on space travel than any other publication in the world. I can think of no more exciting story it could publish, or that I could write, than to be able to report that I have finally found a UFO case that defies any possible prosaic explanation. I would expect to win a Pulitzer Prize, a giant bonus, and great fame. So far, I've had no luck. But who knows, perhaps tomorrow—or next week.

CROP CIRCLES: MESSAGES FROM OUTER SPACE

Freddy Silva

No one has been able to satisfactorily explain the origin of crop circles—circular and geometric patterns in fields that are made by flattening the fields' crops. Freddy Silva discusses research that indicates that the creation of the crop circles causes changes in the plants' cells, the fields' soil, and other physical and chemical anomalies. Silva suggests that these circles may be formed by extraterrestrials as messages intended to make humans aware of their presence. Silva is a researcher and lecturer who has studied crop circles in England.

Crop circles are not a modern phenomenon. They are mentioned in texts as far back as 1678, and almost 200 reports, including eyewitness accounts, have been reported up to 1970. It was first witnessed during modern times in 1972 by Arthur Shuttlewood and Bryce Bond, who had been sitting on the slope of Star Hill near Warminster, England, hoping to catch a glimpse of the strange unidentified flying craft that had made this part of England a UFO Mecca for almost a decade. But what they witnessed on that moonlit night was something more extra-ordinary: a hundred feet away they saw an imprint take shape, a large circular area of plants that collapsed like a lady opening a fan. Since then some eighty eyewitnesses from as far away as British Columbia have reported the formation of crop circles, which occur in under twenty seconds, often accompanied by sightings of unusual balls of light, shafts of light or structured flying craft.

The designs appeared primarily as simple circles and variations on the Celtic cross up into the mid-1980s. Then they developed straight lines and created pictograms. After 1990 they exploded exponentially, and today it is not unusual to come across designs mimicking computer fractals, some occupying areas as large as 200,000 sq. feet. To date there have been over 9,000 reported and documented crop circles throughout the world, with some 90 percent emerging from England. While many still go unreported each year, the emergence of the phenomenon in the world media and the internet has allowed more reports to be lodged.

Reprinted, with permission, from "A Crash Course on Crop Circles for Beginners," by Freddy Silva, *Crop Circular*, no date, published at http://home.clara.net/lovely/education.html.

The Hoax Theories

If you happen to buy the story that all crop circles were originated by two sexagenarians with planks of wood, string and a Ouija board, you are not in the minority. Once in a while, governments like to control public interest in unexplained phenomena by generating a disinformation method called 'debunking', a technique invented during the Cold War for the sad purpose of controlling mass opinion in the face of unexplainable phenomena. . . . This method is very effective because the media provides little or no scientific or factual data with which the public can form an educated opinion on the subject. This absence of evidence is then replaced by ridiculing the subject through association with other 'fringe' topics; so-called experts are brought in to explain away all the events as freak weather conditions or the work of general pranksters or sexually excited animals.

According to TV documentaries, all crop circles up to 1992 were made by two simple, elderly men called Doug and Dave. It has since been discovered through undercover work by researchers such as George Wingfield and Armen Victorian that the D&D story was tied to the British Ministry of Defense (M.O.D.)—in collusion with the CIA, among others. Evidence supplied by a high-ranking informant in the M.O.D. suggested that the government had every intent to discredit the phenomenon by putting forward two hoaxers in an effort to quell growing public interest in crop circles. When confronted to provide evidence on certain claimed formations, Doug and Dave changed their story and admitted that "no, we never made that one" or would simply remain silent when asked to explain the list of features found in the genuine phenomenon; when they claimed making all the formations around the English county of Hampshire it was pointed out that half the known formations had occurred in another county—"Er, no, we didn't do those either," they replied. In the end, not even Doug and Dave knew which ones they had made. And although they claim to have made hoaxes since 1978—the published date of the first design—evidence withheld confirms crop circles dating back into the 1930s. The public has never heard these retractions, nor been given the opportunity to compare the mess created by D&D with the mathematical symmetry of the real phenomenon.

In 1998, however, the surviving member of the deceptive duo did make the incredible admission to British newspapers that he'd been guided by an unknown force.

Since Doug and Dave's inauguration many copycat hoaxers have appeared on the scene. Some do it to disprove or derail researchers, some for profit, some because they are sociopaths, some because they genuinely believe they can communicate back to the phenomenon (with very interesting results, I may add). Prior to 1989 the hoaxing problem was virtually unheard of. After 1990 designs of man-made origin vary by year—in 1992 and 1998 it was as high as 90 percent, in 1996 as low as 20 percent.

That people with a good amount of training can go into a field and eventually create a coherent pattern has never been the issue—recently, a group of known hoaxers was paid to go to conveniently out-of-the-way New Zealand to make an elaborate formation for NBC. . . .

Characteristics of Genuine Crop Circles

The issue is that no man-made crop circle has satisfactorily replicated the features associated with the real phenomenon which has baffled scientists and researchers. Worldwide, some 4,000 crop circles have been created by a force totally at odds with modern science. Central to the hoax angle is that a physical object is required to flatten the crop to the ground, resulting in the breaking of the plant stems. In genuine formations the stems are not broken but bent, normally about an inch off the ground at the plant's first node. The plants are subjected to a short and intense burst of heat which softens the stems to drop just above the ground at 90 degree angles, where they reharden into their new and very permanent position without damaging the plants. Plant biologists are baffled by this phenomenon and farmers, who know how the land ticks, are baffled by this. It is the singlemost method of identifying the real phenomenon. Research and laboratory tests suggest that microwave or ultrasound may be the only method capable of producing such an effect.

Crop circles are sometimes accompanied by trilling sounds, since captured on tape and analysed by NASA as artificial in origin.

Other features that cannot be replicated by hoaxes are the plants' expanded epidermal walls, and drastically extended node bends in fresh formations; also observed are distortions of seed embryos, and the creation of expulsion cavities in the plants as if they have been heated from the inside.

In genuine formations there is also a disruption of comparative analysis of the plant's crystalline structure. Yet in all cases, the plants are not damaged and will continue to grow and ripen if left untouched. This would not be possible had they been trampled by force.

Genuine crop circles are areas of gently laid and swirled plants which create a floor in the same spiralled logarithmic proportions as the Fibonacci Series or Golden Mean, the vortex nature uses to create precision organisms such as shells, sunflowers, the bones on the human hand and galaxies; the floor of crop circles can have up to five layers of weaving, all in counterflow to each other, with every seed head intact and placed beside each other as if arranged in a museum case; the centres can either contain nested, woven, crested, or wreathed swatches of plants—sometimes the center will consist of a single standing plant. They are not perfectly round but slightly oval (a hoax, requiring a fixed central rope, cannot achieve this adequately). Their edges are crisply defined from the flattened crop as if drawn with a compass and incised with surgical precision. Hoaxes, by comparison, bear a stylistic resemblance to tufts of greasy, uncombed hair—and, of course, all their plants have been trampled, bruised and crushed.

Other Anomalies

Other anomalies indicate the ability of the Circlemakers to increase infrared output within and around a new formation, indicating that both the heat content of the plants and the watershed have been affected. Evidence of four non-naturally occurring, short-life radioactive isotopes in the soil inside genuine crop circles has been detected (these dissipate after three or four hours), and the soil in and around them appears to have been baked.

Mathematically, genuine crop circles encode obscure theorems based on Euclidian geometry as well as the unalterable principles of sacred geometry. They have the capacity to alter the local electromagnetic field so that compasses cannot locate north, cameras, cellular phones and batteries fail to operate, and aircraft equipment fails whilst flying over formations. Then there are the Geiger counters recording levels of background radiation up to 300 percent above normal, radio frequencies falling dramatically within their perimeters, animals in local farms avoiding that particular area of the field or simply acting agitated hours before one materializes, and car batteries in entire villages failing to operate the morning after one is found nearby. In some of the major events, local power outages are reported.

Genuine formations also materialize at crossing points along the Earth's magnetic energy currents, influencing the energy pattern of local prehistoric sites. They often reference local Neolithic features in size/shape/direction, and are dowsable upon entry, with as many as 150 concentric rings of energy outside their physical perimeter. In fact, a year after they have been harvested and the field ploughed and re-sown, the energy imprint of the formations will still be dowsed in the same location, long after its physical traces have vanished.

This area of research has allowed for the possibility of crop circles as a healing force, and they are already being successfully employed in resonance therapy around the world, either using people or environments in distress.

They are generally formed at night between the hours of 11:30 P.M. and 4 A.M., traditionally during the shortest evenings of the English year when darkness lasts but four hours, in fields eagerly watched by farmers, military, laser alarms, scientists or hundreds of enthusiasts in their sleeping bags hoping to be the lucky ones to witness a crop circle forming. Some of those lucky few have witnessed large balls of brilliant colour project a beam of golden light into a field which next morning displays a new crop circle. Yet despite many stakeouts and fields rigged with top surveillance equipment, crop circles have appeared out of the mist right under the noses of those looking for them. On one occasion, the Circlemakers even materialized in full view of the British Prime Minister's heavily-guarded country residence.

At Stonehenge in 1996, a pilot reported seeing nothing while flying above the monument, yet 45 minutes later this huge 900-foot formation resembling the Julia Set computer fractal, comprising 145 meticulously laid circles, lay beside the heavily guarded monument. It took a team of 11—including myself—no less than five hours just to survey this formation.

Still not convinced? This web site [http://home.clara.net/lovely/homepg.html] contains a sampling of the on-going research dedicated to enlightening the public. More will be added as time goes by. Look at the pictures, study the research or better still, visit a genuine crop circle. You'll get the message pretty quickly.

And when you do, tell this story to a friend.

CROP CIRCLES ARE HOAXES

Matt Ridley

Science writer Matt Ridley writes in the following viewpoint that most crop circles are undoubtedly man-made rather than evidence of alien phenomena. Ridley himself has made a few crop circles on farms in England and describes the process. Nevertheless, he writes, gullible reporters continue to accept the idea that crop circles are not simply the work of pranksters. Ridley concludes that more skepticism is warranted when examining the conclusions of self-proclaimed crop circle experts. Ridley is the author of *Genome: The Autopsy of a Species in 23 Chapters.*

Laurance Rockefeller, an American billionaire who is also a UFO nut, has been persuaded to part with some of his billions to "carry out aerial research" into crop circles in southwest England. Is there no end to the gullibility of the wealthy?

Man-Made Phenomena

I made my first crop circle in 1991. My motive was to prove how easy they were to make, because I was convinced that crop circles were all man-made. It was the only explanation nobody seemed interested in testing.

Late one August night, with one accomplice—my brother-in-law from Texas—I stepped into a field of nearly ripe wheat in northern England, anchored a rope into the ground with a spike and began walking in a circle with the rope held near the ground. It did not work very well: the rope rode up over the wheat. But with a bit of help from our feet to hold down the rope, we soon had a respectable circle of flattened wheat.

Two days later, there was an excited call from the farmer. I had fooled my first victim. I subsequently made two more crop circles using much superior techniques. A light garden roller, designed to be filled with water, proved helpful. Eventually, I hit on the technique used by the original circle-makers—plank-walking, or pushing down the crop with a plank suspended from two ropes. To make the depression circular is a simple matter of keeping an anchored rope taut.

Reprinted, with permission, from "Comment: Pranksters Run Rings Around the Gullible," by Matt Ridley, *Daily Telegraph*, May 24, 1999.

Getting into the crop without leaving traces is a lot easier than is usually claimed. In dry weather, and stepping carefully, you can leave no footprints or tracks at all. There are other ways of getting into the crop even more stealthily. One group of circle-makers use two tall bar stools and jump from one to another. I soon found I could make a sophisticated pattern with very neat edges and no tracks in less than an hour.

The Science of "Cereology"

But to my astonishment, throughout the early 1990s, television and newspaper reporters continued to say that it was impossible that all crop circles could be man-made. They cited "cereology" experts to this effect and never checked for themselves. There were said to be too many circles to be the work of a few "hoaxers" (but this assumed that each took many hours to make); or that circles appeared in well-watched crops (simply not true); or that the creation of circles was accompanied by unearthly noises (when these were played back, even I recognised the nocturnal song of the grasshopper warbler).

The most ludicrous assertion was that "genuine" circles could be distinguished from "hoaxed" ones by "experts". Even after the chief self-styled "expert", Terence Meaden, asserted on camera that a circle was genuine when it had been filmed in the making by Channel 4, the programme let him off the hook, by saying he might just have made a mistake this time.

I soon met other crop circle-makers, though I was never lucky enough to encounter the two who started it: Doug Bower and Dave Chorley. People such as Robin Allen of Southampton University and Jim Schnabel, author of *Round in Circles,* also found it all too easy to fool the self-appointed experts, but all too hard to dent the gullibility of reporters. When Bower and Chorley confessed, they were denounced on television as frauds. My own newspaper articles were dismissed as "government disinformation" and it was hinted that I was in MI5 [an agency of the British government], which was flattering (and untrue).

The whole episode taught me two important lessons I have tried to employ ever since. First, treat all experts with scepticism and look out for their vested interests—many "cereologists" made a lot of money from books, and week-long tours of crop circles are offered to new-age Americans for more than $2,000 per person. Second, never underestimate the gullibility of the media. Even *Science* magazine and the *Wall Street Journal* published articles that failed to take the man-made explanation seriously.

As for who made the complicated mathematical and fractal patterns that appeared in the mid-1990s, I have no idea. But Occam's razor suggests it was more likely to be undergraduates than aliens.

A GIFT FROM THE ALIENS

Patrick David MacKondy

In the following selection, Patrick David MacKondy recounts how he was abducted twice by aliens, first when he was six years old and again when he was twenty-six. He believes that the aliens placed something into his body during his first abduction and surgically removed it during his second abduction, leaving him with several noticeable scars. According to MacKondy, the aliens also gave him special gifts—extrasensory perception and the ability to communicate with animals—as a "thank you" for letting them experiment with his body. However, his abduction experiences also have had some drawbacks, such as a new-found fear of planes and of doctors and dentists. MacKondy is a creative director for a cable television advertising company in Ohio.

Twinsburg, Ohio, was a great place to raise a family in 1966. I was a dapper, sandy-haired six-year-old. I lived in a beautiful new apartment complex with my father, mother, and sister. My father was an advertising director for a regional grocery chain, and he was in the process of building us a stately Colonial house down the street. I have wonderful memories of waiting for my dad to get home from work at 5:30. We would walk to our future home to check on the day's progress. Even today, when I smell freshly cut plywood or drywall, memories of those days rush back to me.

I eagerly anticipated our new home's completion. It was only two blocks from our apartment, but I would feel a lot safer once we moved. For even though I had a wonderful childhood, there was one experience in that apartment that I will never forget.

Our next-door neighbors were Dutch. They were excited because their dearest friends from Holland were moving to the United States and were going to rent the apartment across the street. They said that I would have a new playmate named Lazette. I wasn't crazy about playing with a girl, and hoped we would move before she arrived.

My wishful thinking came to naught, however. Lazette and her parents moved in, and that summer she drove me crazy by trying to be one of the boys.

Reprinted, with permission, from "What the Aliens Left Me," by Patrick David MacKondy, *Fate*, September 1998.

A Frightening Experience

One afternoon Lazette and her parents begged my parents to let me spend the night at their apartment. This would be the first time that I had ever slept over at anyone's house. My parents, as overprotective as they were, gave in.

We went to the movies, out to eat, and back to their apartment. I put my pajamas on and went to bed in Lazette's bedroom. Strangely, Lazette slept between her parents in their room, and left me all alone.

Later that night, I woke up to find both of her parents looking at me. Their noses were only inches from mine, and I was frightened. I must have passed out or fallen back to sleep. When I awoke again, I was outside, alone, looking up into the sky at a bright beam of light that must have been 10 or 12 feet in diameter. I watched the light shrink until it went back up into the sky.

I heard a male voice in my mind. "We will see you again someday, when you are a lot older," it said.

Lazette's parents walked me back to the apartment, where I fell fast asleep. I was crying hard and wanted to go home. They were laughing. I thought I would never see my parents again.

The next day, I told my parents all about it. They said that the rich food I had eaten must have given me a nightmare. I knew it had been real, though, because Lazette's parents had been outside in their pajamas, and that morning I had seen mud on Lazette's mother's slippers while she prepared my breakfast.

Time passed and I tried to put the memory of that night out of my head. I knew that something strange had happened to me, but I didn't know what. As I got older I told myself that it had been one heck of a nightmare. But something deep inside me remained unsettled.

Moving Onward

From Twinsburg our family moved to Youngstown, Ohio. My father had accepted a position as the advertising director for another grocery chain. I became a happy, successful adult. I began modeling men's clothing for a local department store, and quickly progressed to photo shoots in New York City, Miami, Los Angeles, and other locations.

Then, when I was 26, I had an experience that brought back the memory of my strange evening in Twinsburg.

I was going to bed early on a muggy night in August. I had to get up in the morning for a catalog shoot, and I wanted to look rested. Alone in my bedroom, I was scared out of my mind for no apparent reason. I had an inexplicable feeling that I was going to get a visit from someone—or something.

I was lying quietly in my bed when a small, transparent ball of light entered my room. The sphere shot at my face and buzzed around my nose like a bee. I tried to smack it, but my hand went right through it.

The more I tried to get it away from my face, the more tired I grew, and I eventually fell asleep. It was as though the light was sent into my room to drain me of my strength.

The next morning I awoke tired and weak. I thought I was coming down with a flu. I showered and got ready for my catalog assignment, and off I went.

When I arrived at the shoot I was given five bathing suits. After I changed into the first one, the owner of the swimwear company approached me and started to examine parts of my body. She said that she didn't remember so many scars. I looked at where she was pointing and got sick to my stomach. I had at least fifteen scars all over my body. They were about four inches long and one-sixteenth inch wide. I had never seen them before, but they appeared to be at least a year old. Makeup covered them nicely, however, and I was permitted to continue the session.

My route home from the shoot took me close to Twinsburg, and something told me to drive by the old apartment building. A nostalgic pang went through me as I pulled up in front.

Then, passing Lazette's apartment, I had a flashback, clearer than ever before, of the night I spent there. I saw an image of the beam of light, in greater detail than I had previously remembered, and realized it had come from a ship. I recalled seeing a row of colored lights in hues that didn't exist in our color spectrum.

I realized that I had been to another planet, once as a six-year-old child, and again the previous night. I opened up the car door and vomited.

As I sat in my car, more information poured into my brain, which started to sort the data and make sense out of my experiences. During the drive home I put things into perspective. I was not scared. I knew I would not be abducted again, that the aliens (or "visitors," as I now call them) were done with me.

I felt as though I had been given new insights, power, love, and a greater sense of the universe. I also felt cleansed—what the visitors put into me at age six, they removed 20 years later.

What They Left Me

The scars proved to me that I had been involved in an alien operation, but the visitors left me with no memory of it, knowing it would be too painful to live with. As a thank-you for letting them use my body, I guess, they left me with gifts that are definitely out of this world.

I learned about my first gift a year after my second encounter. My friends and I were in a large nightclub in Austintown, Ohio. It was a rowdy place where fights often erupted. The club was packed tight that evening, and my buddies and I stood pressed up against the bar.

Suddenly I heard a humming noise, and without thinking, I jumped up and spun around, sitting smack down on the bar. A split-second later, a 275-pound guy smashed a beer bottle over the head of the guy standing next to him.

The fight was horrifying. If I hadn't moved, I would have been killed. My friends must have asked me at least 50 times what made me jump out of the way. There had been no sign of a fight. It was then I knew I had a gift.

Two months later in the same club, a group of us were talking, and I finished a woman's sentence for her. It scared her to death. She says I'm a warlock or that I'm not of this Earth, and she still won't talk to me.

Things became even stranger about a week later. I would start to hear a humming sound in my head just before the phone would ring. Now it happens every time. I say out loud: "Someone's calling." And, sure enough, the phone rings.

Another Gift

I found out about my second gift during the summer of 1990. I was outside and saw a strange dog in our yard. I looked at the dog and thought very hard: *Come here, pup. Come here, pup. Don't be afraid. Come and see me.* The dog immediately came to see me.

I thought it was just a coincidence, so I tried something else: *Give me your paw,* I thought. The dog did.

The next day a neighbor was walking his dog, a friendly poodle who had never growled in his life. We started talking about cars, as I had just bought a new Camaro. He came down the driveway to take a closer look at my car, and I thought I would test my abilities again.

As I was showing him the engine, I looked down at the poodle, then back at him. I envisioned myself hitting the man. The dog immediately growled and tried to bite me. My neighbor could not believe what had happened. He thought his poodle was nuts. I have been communicating with animals ever since.

My most recent encounter was at a local street fair where there was a wild animal exhibit. I approached a lion cage and asked the owner if I could pet his three-year-old lion. The man laughed like Santa Claus. In a deep, jolly voice he told me that I would be chewed up alive. He said that he was the only one who could touch his lions.

I chuckled to myself and waited for him to take a break. When he did I snuck behind the cage bars and cooed the lion to sleep. My girlfriend and I still laugh about the look on the owner's face when he returned to find me in the cage with his fierce lion.

Unearthly Premonitions

In the winter of 1997, I began having beautiful dreams. These dreams were in vivid color and often foretold the next day's events. Once, I dreamt about an old neighbor whom I had not seen in over five years. The following day there was a knock at the door. It was her.

Three weeks later I dreamt that the person to whom I had sold my old car stopped at my house to tell me that it was still running like a clock. The next day, I saw the car pull into my driveway.

And it wasn't more than a week later that I had a dream that my aunt was sick, and was asking for my family's help. I also dreamt that my father was sharing my dream. Not only did my aunt call the next day, but my father had also had the same dream the night before.

A Mixed Blessing

In addition to my gifts, I also have a few phobias that I attribute to my abductions. The first is a severe fear of flying. It's not the height so much as it is being contained in a big steel vehicle, knowing I will not land for a couple hours. I think this phobia comes from my ride in the visitors' ship.

I also fear visiting doctors or dentists. My last dental appointment was embarrassing. The taste of the metal object probing my mouth sent me flying out of the chair and down the hallway. I believe this is connected to the probing and scarring I received from the visitors.

Before August 1987, I had no fear of flying or of visiting the dentist. In fact, I had flown more than 40 times, as my father had owned a travel agency and our family had flown practically for free.

Many people ask why I don't visit a psychiatrist specializing in regressions, or why I didn't report my encounter to NASA. I tell them that I really don't care to revisit my experiences, nor do I feel a need to come forward when the government already knows that these abductions occur. I wrote this article to share my story with people who may have experiences similar to mine.

I warned my family I had written this article and asked them not to think that I was crazy. Even though they knew about my abilities, my family told me that discussing them would make people think I was out of my mind.

As I read them the first paragraphs, a small tear ran down my father's cheek. Choked with emotion, dad told us that while living in the apartments, he had had a nightmare way back in 1966, one he was never able to get out of his mind. In his horrifying, realistic dream, Dad saw a bright light outside his bedroom window. He dreamt that he got out of bed and pulled back the curtains to see his six-year-old son, in his pajamas, being led back to the apartment by Lazette's parents as a large ship disappeared into the sky.

Abducted by Aliens

Laura

The following selection is by an anonymous author who identifies herself only as Laura. She claims that she has seen aliens and been abducted by them since she was a young girl. She learned to keep such experiences a secret from others, she writes, because she would often be ridiculed or punished if she talked about them. Laura maintains that as she matured and entered adulthood, the aliens started to conduct gynecological procedures on her. She began to use alcohol as a way to obliterate her fear, but eventually she realized that she had to regain control of her life. Laura relates how she joined UFO support groups where her experiences were validated, and she reports that she is now beginning to feel secure and hopeful about her life.

From my earliest memory I knew beyond a doubt that my life was different than other children's lives. I feared the visitors that would come to my bedroom unseen by my parents in the night. At the time that I was a child in the late 1940s there were no movies about *ET*, *Close Encounters*, or *Independence Day*. I knew, however, that I was fearful of "space aliens" who came to visit when I was alone and defenseless. Their preferred method of entry to my room was through a window and my parents were perplexed at my inordinate fear of sleeping near one. They would also lure me outdoors where they would be waiting. I struggled against these liaisons but in the end their will prevailed. I would hide in a secret refuge in a walk-through closet that led to my parents' room. I would see bright lights emanating from my room that no one else was awake to see.

Childhood

In childhood, and to some degree in the present, I had an overwhelming fear of doctors and dentists with their bright lights and sharp tools. Fear was my constant companion—fear of the darkness, of windows, of being alone in a church, the outdoors, and of parked aircraft—that I imagined would come to life along with the occupants.

Reprinted, with permission, from "Memoirs of Shadows," by Laura, published on the International Center for Abduction Research website at www.ufoabduction.com/ abductees.html.

I learned early on not to admit my true fears to adults as they would very quickly tell me that space aliens did not exist. If I persisted I was punished. Of the many events that occurred in my childhood two events stand out—one of which was explored through the use of regression. The first one is a partial memory of encountering a male and female alien couple when I was approximately seven years of age. One night I was coming home from visiting my grandmother who lived nearly adjacent to us. As I reached the halfway point I saw two figures standing behind the fence. My heart began to race as these were my night visitors—not outwardly different from one another but one male and one female. They did not speak aloud but they called me to come to them. In horror I shouted I would not and tried to run the rest of the way home calling them monsters. The next thing they uttered I have not forgotten to this day. They told me that they were my "real parents" and that I should go with them. I have little memory of what happened after that; however, I like to think that I ran and made it home.

My next clear memory was of seeing the ghost of my grandmother shortly after her death when I was twelve. This memory was further explored with the use of hypnosis and proved to be far different than the screen memory I had of the event. In the morning my family and I had attended my grandmother's funeral and then all the relatives went to my grandparents' home for a gathering. I remember feeling uncomfortable with all of the emotional and grieving adults and I went into a bedroom to get away from them and perhaps take a nap. I entered the room and shut the door behind myself but in a few seconds I realized I was no longer alone. In the corner of the room was a floating figure that I thought was my grandmother's ghost. Upon clarifying the memory I was able to see that the figure appeared far different than my grandmother—no hair, large black eyes that slanted, very thin arms, and pale whitish-gray skin. Frightened, I clung desperately to the door handle in hopes of escape. (I had always thought that I had fled after seeing that ghostly figure but that proved to be untrue.) The figure moved toward me and stared at me with enormous black eyes. It told me to come with it. I initially objected and resisted, but I uncontrollably went with the alien. During the regression the saddest moment was when I realized that I no longer had my hand on the door handle and there was no possibility of escape.

I was taken to a small waiting UFO and had procedures performed that included a sexual-gynecological procedure at my young age of twelve. I was returned to the bedroom and forgot most of the details other than I saw something frightening that I believed was a ghost. When I found my mother she indicated angrily that people were looking for me and that I was upsetting everyone with my talk of ghosts. Events of this type were interspersed throughout my childhood— sometimes clear and sometimes not at all.

Adolescence/Teen Years

Adolescence was for me, as everyone, an awkward age. Not quite grown up but flying towards it. The fear remained but I could no longer run to adults for comfort and protection. I was developing feelings about my sexuality but most of the time I was still a child. To the grays I was of reproductive age and of use to them. Teen years are always turbulent and mine were as well, but the added hidden dimension was ever present. Odd occurrences were the norm for me. I worked hard at school, but nights were unpredictable. I tried in my own way to create a sense of safety. In my teen years I found that alcohol often quickly obliterated my fear. This would prove to be a destructive pattern for my later years. It was during these years of change that I began to feel that I was leading a double life—being a young teen by day, and a reproductive age female when I was abducted. As I got older I began to rely on alcohol at night to ease the fear I always felt after dark. In my teen years I was not abducted as frequently as in adulthood but still enough to maintain that gnawing fear of being taken.

Adulthood

When I became an adult I had already had many years of training and indoctrination in what it was like to be a victim of an act that was both unreportable and unbelievable. I continued to have abductions but I had the comfort of two things—alcohol and the "spiritual" event people. I became involved with some New Age and traditional spiritual practices and they for the most part were able to explain my experiences as visitations from angels, spirit guides, demons, or ghosts. The problem was that these explanations didn't always work as the grays couldn't always pull off looking like any of these. My screen memories were somewhat convincing in appearance, but the behavior of these entities never quite fit. Why do spirit guides need to perform gynecological procedures and why couldn't I remember all the great stuff they were supposedly teaching me? Why do angels and demons need spaceships? In my twenties I was driving down a highway when "the devil" appeared in the back seat of my car and tried to take it over. Also several times the spirit of my deceased grandfather appeared to me—always when I was alone. These events have not been fully explored but I suspect they are not paranormal events.

The fear remained and my alcohol consumption eventually had to stop. The spiritual solution quit working, so I turned to the UFO groups. I joined MUFON [Mutual UFO Network] and it was there that I met Dr. David Jacobs. At that point in my life I was fairly certain of my abductions and needed answers and support. I was more aware of the abductions and some of the memories were very detailed and I turned to Dr. Jacobs for help investigating them. Regression hypnosis is not to be undertaken lightly and it tends to open a door that can never be closed again. I do not regret my decision to proceed and it gave me a new peace that I cannot fully explain. It also presented new problems, but they are not shadowy and fragmentary. I have had much support with the memories and ongoing events. The events have been frequent but I feel somehow more able to deal with them. I have a video camera trained on me at night. It has provided relief as I perceive a decrease in events and a feeling of security that has allowed me to sleep with less fear. Nothing works 100%, but I am willing to take something that works perhaps 80% of the time.

I feel at times that I am leading two lives, one that I share with everyone and the other one that involves abductions. As I am new to these awarenesses I am still learning how to integrate all of this information. The grays are not spirit guides and I believe not here to help anyone but themselves. As Dr. Jacobs once told me, I have broken through their secrecy and now I feel I have hope.

ABDUCTION BY ALIENS OR SLEEP PARALYSIS?

Susan Blackmore

Susan Blackmore is a psychology professor at the University of West England and a fellow at the Committee for the Scientific Investigation of Claims of the Paranormal. In the following selection, she posits a variety of psychological explanations that can account for the memories and sensations most often described by alien abductees. One such explanation is sleep paralysis, in which a person wakes up paralyzed in the night and experiences auditory and visual hallucinations. Blackmore criticizes a 1992 survey that claimed that millions of Americans have been abducted by aliens. She cites an experiment of her own that points to sleep paralysis as a more plausible explanation for the abduction memories the survey describes.

If you believe one set of claims, nearly four million Americans have been abducted by aliens. This figure has been widely publicized and is often assumed to mean that millions of people have been visited by members of an alien species and, in some cases, physically taken from their beds, cars, or homes to an alien craft or planet.

Personal accounts of abduction by aliens have increased since the publication of Budd Hopkins's books *Missing Time* (1981) and *Intruders* (1987) and Whitley Strieber's *Communion* (1987). There is considerable variation among the accounts, but many fit a common pattern. [Dan] Wright (1994) summarized 317 transcripts of hypnosis sessions and interviews from 95 separate cases and concluded, "Numerous entity types have been visiting our planet with some regularity." However, the "gray" is clearly the most common alien and over the years a typical account has emerged.

The experience begins most often when the person is at home in bed and most often at night, though sometimes abductions occur from a car or outdoors. There is an intense blue or white light, a buzzing or humming sound, anxiety or fear, and the sense of an unexplained presence. A craft with flashing lights is seen and the person is transported

Reprinted, with permission, from "Abduction by Aliens or Sleep Paralysis?" by Susan Blackmore, *Skeptical Inquirer*, May 1998.

or "floated" into it. Once inside the craft, the person may be subjected to various medical procedures, often involving the removal of eggs or sperm and the implantation of a small object in the nose or elsewhere. Communication with the aliens is usually by telepathy. The abductee feels helpless and is often restrained, or partially or completely paralyzed.

The "gray" is about four feet high, with a slender body and neck, a large head, and huge, black, slanted, almond-shaped eyes. Grays usually have no hair and often only three fingers on each hand. Rarer aliens include green or blue types, the taller fair-haired Nordics, and human types who are sometimes seen working with the grays.

The aliens' purpose in abducting Earthlings varies from benign warnings of impending ecological catastrophe to a vast alien breeding program, necessitating the removal of eggs and sperm from humans in order to produce half-alien, half-human creatures. Some abductees claim to have seen fetuses in special jars, and some claim they were made to play with or care for the half-human children.

Occasionally, people claim to be snatched from public places, with witnesses, or even in groups. This provides the potential for independent corroboration, but physical evidence is extremely rare. A few examples of stained clothing have been brought back; and some of the implants have reportedly been removed from abductees' bodies, but they usually mysteriously disappear.

Theories

How can we explain these experiences? Some abductees recall their experiences spontaneously, but some only "remember" in therapy, support groups, or under hypnosis. We know that memories can be changed and even completely created with hypnosis, peer pressure, and repeated questioning. Are "memories" of abduction created this way? Most of Wright's ninety-five abductees were hypnotized and/or interviewed many times. Hopkins is well known for his hypnotic techniques for eliciting abduction reports, and [John] Mack also uses hypnosis. However, there are many reports of conscious recall of abduction without hypnosis or multiple interviews, and the significance of the role of false memory is still not clear.

Another theory is that abductees are mentally ill. This receives little or no support from the literature. [Ted] Bloecher, [Aphrodite] Clamar, and Hopkins (1985) found above-average intelligence and no signs of serious pathology among nine abductees, and [June] Parnell (1988) found no evidence of psychopathology among 225 individuals who reported having seen a UFO (although not having been abducted). [Nicholas P.] Spanos et al. (1993) compared forty-nine UFO reporters with two control groups and found they were no less intelligent, no more fantasy prone, and no more hypnotizable than the controls. Nor did they show more signs of psychopathology. They did, however, believe more strongly in alien visitations, suggesting that such beliefs allow people to shape ambiguous information, diffuse physical sensations, and vivid imaginings into realistic alien encounters.

Temporal lobe lability has also been implicated. People with relatively labile temporal lobes are more prone to fantasy, and more likely to report mystical and out-of-body experiences, visions, and psychic experiences. However, Spanos et al. found no difference in a temporal lobe lability scale between their UFO reporters and control groups. M. Cox (1995) compared a group of twelve British abductees with both a matched control group and a student control group and, again, found no differences on the temporal lobe lability scale. Like Spanos's subjects, the abductees were more often believers in alien visitations than were the controls.

A final theory is that abductions are elaborations of sleep paralysis, in which a person is apparently able to hear and see and feels perfectly awake, but cannot move. The International Classification of Sleep Disorders reports that sleep paralysis is common among narcoleptics, in whom the paralysis usually occurs at sleep onset; is frequent in about 3 to 6 percent of the rest of the population; and occurs occasionally as "isolated sleep paralysis" in 40 to 50 percent. Other estimates for the incidence of isolated sleep paralysis include those from Japan (40 percent), Nigeria (44 percent), Hong Kong (37 percent), Canada (21 percent), Newfoundland (62 percent), and England (46 percent).

The Sleep-Paralysis Experience

In a typical sleep-paralysis episode, a person wakes up paralyzed, senses a presence in the room, feels fear or even terror, and may hear buzzing and humming noises or see strange lights. A visible or invisible entity may even sit on their chest, shaking, strangling, or prodding them. Attempts to fight the paralysis are usually unsuccessful. It is reputedly more effective to relax or try to move just the eyes or a single finger or toe. Descriptions of sleep paralysis are given in many of the references already cited and in [David J.] Hufford's (1982) classic work on the "Old Hag." I and a colleague are building up a case collection and have reported our preliminary findings.

Sleep paralysis is thought to underlie common myths such as witch or hag riding in England, the Old Hag of Newfoundland, Kanashibari in Japan, Kokma in St. Lucia, and the Popobawa in Zanzibar, among others. Perhaps alien abduction is our modern sleep paralysis myth.

Spanos et al. (1993) have pointed out the similarities between abductions and sleep paralysis. The majority of the abduction experiences they studied occurred at night, and almost 60 percent of the "intense" reports were sleep related. Of the intense experiences, nearly a quarter involved symptoms similar to sleep paralysis.

Cox (1995) divided his twelve abductees into six daytime and six nighttime abductions and, even with such small groups, found that the nighttime abductees reported significantly more frequent sleep paralysis than either of the control groups.

I suggest that the best explanation for many abduction experiences is that they are elaborations of the experience of sleep paralysis.

Imagine the following scenario: A woman wakes in the night with a strong sense that someone or something is in the room. She tries to move but finds she is completely paralyzed except for her eyes. She sees strange lights, hears a buzzing or humming sound, and feels a vibration in the bed. If she knows about sleep paralysis, she will recognize it instantly, but most people do not. So what is she going to think? I suggest that, if she has watched TV programs about abductions or read about them, she may begin to think of aliens. And in this borderline sleep state, the imagined alien will seem extremely real. This alone may be enough to create the conviction of having been abducted. Hypnosis could make the memories of this real experience (but not real abduction) completely convincing.

The Roper Poll

The claim that 3.7 million Americans have been abducted was based on a Roper Poll conducted between July and September 1991 and published in 1992. The authors were Budd Hopkins, a painter and sculptor; David Jacobs, a historian; and Ron Westrum, a sociologist. In its introduction John Mack, professor of psychiatry at Harvard Medical School, claimed that hundreds of thousands of American men, women, and children may have experienced UFO abductions and that many of them suffered from distress when mental health professionals tried to fit their experiences into familiar psychiatric categories. Clinicians, he said, should learn "to recognize the most common symptoms and indications in the patient or client's history that they are dealing with an abduction case." These indications included seeing lights, waking up paralyzed with a sense of presence, and experiences of flying and missing time. The report was published privately and mailed to nearly one hundred thousand psychiatrists, psychologists, and other mental health professionals encouraging them to "be open to the possibility that something exists or is happening to their clients which, in our traditional Western framework, cannot or should not be."

The Roper Organization provides a service for other questions to be tacked on to their own regular polls. In this case, 5,947 adults (a representative sample) were given a card listing eleven experiences and were asked to say whether each had happened to them more than twice, once or twice, or never. The experiences (and percentage of respondents reporting having had the experience at least once) included: seeing a ghost (11 percent), seeing and dreaming about UFOs (7 percent and 5 percent), and leaving the body (14 percent). Most important were the five "indicator experiences": 1) "Waking up paralyzed with a sense of a strange person or presence or something else in the room" (18 percent);

2) "Feeling that you were actually flying through the air although you didn't know why or how" (10 percent); 3) "Experiencing a period of time of an hour or more, in which you were apparently lost, but you could not remember why, or where you had been" (13 percent); 4) "Seeing unusual lights or balls of light in a room without knowing what was causing them, or where they came from" (8 percent); and 5) "Finding puzzling scars on your body and neither you nor anyone else remembering how you received them or where you got them" (8 percent).

The authors decided that "when a respondent answers 'yes' to at least four of these five indicator questions, there is a strong possibility that individual is a UFO abductee." The only justification given is that Hopkins and Jacobs worked with nearly five hundred abductees over a period of seventeen years. They noticed that many of their abductees reported these experiences and jumped to the conclusion that people who have four or more of the experiences are likely to be abductees.

From there, the stunning conclusion of the Roper Poll was reached. Out of the 5,947 people interviewed, 119 (or 2 percent) had four or five of the indicators. Since the population represented by the sample was 185 million, the total number was 3.7 million—hence the conclusion that nearly four million Americans have been abducted by aliens.

Why did they not simply ask a question like, "Have you ever been abducted by aliens?" They argue that this would not reveal the true extent of abduction experiences since many people only remember them after therapy or hypnosis. If abductions really occur, this argument may be valid. However, the strategy used in the Roper Poll does not solve the problem.

With some exceptions, many scientists have chosen to ignore the poll because it is so obviously flawed. However, because its major claim has received such wide publicity, I decided a little further investigation was worthwhile.

Real Abductions or Sleep Paralysis?

The real issue raised by the Roper Poll is whether the 119 people who reported the indicator experiences had actually been abducted by aliens.

Since the sampling technique appears to be sound and the sample large, we can have confidence in the estimate of 2 percent claiming the experiences. The question is, Have these people really been abducted? The alternative is that they simply have had a number of interesting psychological experiences, the most obviously relevant being sleep paralysis. In this case, the main claim of the Roper Poll must be rejected. How do we find out?

I reasoned that people who have been abducted (whether they consciously recall it or not) should have a better knowledge of the appearance and behavior of aliens than people who have not. This leads to two simple hypotheses.

The Roper Poll assumes that people who have had the indicator experiences have probably been abducted. If this assumption is correct, people who report the indicator experiences should have a better knowledge of what aliens are supposed to look like and what happens during an abduction than people who do not report indicator experiences. If the assumption is not correct, then their knowledge should be no greater than anyone else's—indeed, knowledge of aliens should relate more closely to reading and television-watching habits than to having the indicator experiences if abductions do not really occur.

I decided to test this using both adults and children here in Bristol. It might be argued that genuine abductees wouldn't be able to remember the relevant details so I needed to use a situation that would encourage recall. I decided to relax the subjects and tell them an abduction story, and then ask them to fill in missing details and draw the aliens they had seen in their imagination.

Method

Subjects were 126 school children aged 8 to 13 and 224 first-year undergraduates aged 18 and over. The children came from two schools in Bristol. They were tested in their classrooms in groups of 22 to 28. The first group of 22 children had a slightly different questionnaire from the other groups and, is therefore, excluded from some of the analyses. The adults were psychology and physiotherapy students at the University of the West of England tested in three large groups. The procedure for the children is described below. The procedure was slightly simplified and the story slightly modified for the adults.

I first spent about half an hour talking to the children about psychology and research so that they got used to me. I then asked them to relax—as much as they could in the classroom. Many laid their heads on their desks, some even lay down on the floor. I asked them to imagine they were in bed and being read a bedtime story. I suggested they try to visualize all the details of the story in their minds while I read it to them. I then read, slowly and clearly, a story called "Jackie and the Aliens," in which a girl is visited in bed at night by a strange alien who takes her into a spacecraft, examines her on a table, and brings her back unharmed to bed. The story includes such features as traveling down a corridor into a room, being laid on a table, seeing alien writing, and catching a glimpse of jars on shelves. However, precise details are not given.

At the end of the story, I asked the children to "wake up" slowly and to try to remember as much as they could of the details of the story. I then handed out the questionnaires. Each questionnaire contained five multiple-choice questions about the alien, the room, and table; and the children were asked to describe what was in the jars and to draw the alien writing. There were also six questions based on those in the Roper Poll: Have you ever seen a UFO? Have you ever seen a ghost? Have you ever felt as though you left your body and could fly around without it (an out-of-body experience, or OBE)? Have you ever seen unusual lights or balls of light in a room without knowing what was causing them, or where they came from? Have you ever woken up paralyzed, that is, with the feeling that you could not move? And, Have you ever woken up with the sense that there was a strange person or presence or something else in the room? (Note that in the Roper Poll, the question about paralysis was compounded with the question of the sense of presence. Here, two separate questions were asked. Note also that the last four of these questions were based on the indicator questions from the Roper Poll.) The questions were slightly altered to make them suitable for young children, and I did not ask about scars or missing time. A question about false awakenings (dreaming you have woken up) was also included, and two questions about television-watching habits.

Finally, all groups except one of the adult groups were asked to draw pictures of the alien they had imagined in the story.

Results

Large numbers of both adults and children reported having had most of the experiences. The percentages are shown in Table 1.

Table 1

Experience	Adults	Kids
Ghosts	14%	33%
OBEs	35%	33%
UFOs	8%	28%
False Awakenings	83%	57%
Sleep Paralysis	46%	34%
Presence	68%	56%
Lights	17%	28%

For each person, an "alien score" from 0 to 6 was given for the number of "correct" answers to the questions about the alien (that is, answers that conformed to the popular stereotype), and another score for the number of Roper Poll indicator experiences reported (0–4).

For the children, the mean alien score was 0.95, and the mean number of experiences 1.51. There was no correlation between the two measures ($rs = -0.03$, $n = 101$, $p = 0.78$). The drawings of aliens were roughly categorized by an independent judge into "grays" and "others." Twelve (12 percent) of the children drew grays and 87 did not. Not surprisingly, those who drew a gray also achieved higher alien scores ($t = 3.87$, 97 df, $p < 0.0001$), but they did not report more of the experiences ($t = 0.66$, 95 df, $p = 0.51$).

Those children who drew grays did not report watching more television. Nor was there a correlation between the amount of television watched and the alien score (rs = 0.002, n = 101, p = 0.98). Oddly, there was a small positive correlation between the amount of television watched and the number of experiences reported (rs = 0.25, n = 101, p = 0.01).

For the adults, mean alien score was 1.23 and mean number of experiences 1.64. Again, there was no correlation between the two measures (rs = 0.07, n = 213, p = 0.29). Seventeen of the adults drew grays, and 103 did not. Again those who drew a gray achieved higher alien scores (t = 6.11, 118 df, p < 0.0001) but did not report more experiences (t = 0.14, 115 df, p = 0.89).

Among the adults, those who drew grays were those who watched more television (U = 534, n = 100, 17, p < 0.01), and the amount of television watched correlated positively with the alien score (rs = 0.20, n = 217, p = 0.003).

Discussion

These results provide no evidence that people who reported more of the indicator experiences had a better idea of what an alien should look like or what should happen during an abduction. If real gray aliens are abducting people from Earth, and the Roper Poll is correct in associating the indicator experiences with abduction, then we should expect such a relationship. Its absence in a relatively large sample casts doubt on these premises.

Among the adults (though not the children), there was a correlation between the amount of television they watched and their knowledge about aliens and abductions. This suggests that the popular stereotype is obtained more from television programs than from having been abducted by real aliens.

Our sample certainly included enough people who reported the indicator experiences. Although not all the indicator experiences were included, for the four questions that were used, the incidence was actually higher than that found by the Roper Poll. Presumably, therefore, many of my subjects would have been classified by Hopkins, Jacobs, and Westrum as having been abducted. The results suggest this conclusion would be quite unjustified.

These findings do not and cannot prove that no real abductions are occurring on this planet. What they do show is that knowledge of the appearance and behavior of abducting aliens depends more on how much television a person watches than on how many "indicator experiences" he or she has had. I conclude that the claim of the Roper Poll, that 3.7 million Americans have probably been abducted, is false.

CHAPTER 4

UFOs AND CULTURE

A HISTORICAL EXAMINATION OF UFOS AND ART

Daniela Giordano

In the following essay, Daniela Giordano raises the possibility that unidentified flying objects have been portrayed in art for hundreds of years. She discusses several examples of fifteenth-through nineteenth-century paintings in which the artists have included objects that appear to be UFOs. Aircraft had not yet been invented when these paintings were made, she notes, so the depiction of what appear to be spaceships is especially puzzling. Although the painted objects' resemblance to UFOs is striking, Giordano concludes that it is premature to identify them as spaceships without knowing what the artists actually saw. Giordano won the Fund for UFO Research's 1998 Donald E. Keyhoe Journalism Award for the article this piece was adapted from—"Do UFOs Exist in the History of the Arts?"—which originally appeared in the March 1998 issue of *Nuovo Orione*, an Italian astronomy magazine.

Since the dawn of human history, the sky has inspired our collective imagination. It is no coincidence that astronomy is the most ancient science, nor that its origins are intermingled with the roots of civilization and religion. Humans have always felt a need to reproduce celestial events—first on the walls of caves, and then on canvas. The annals of history, art, archaeology, and anthropology are occasionally rewritten as a result of new discoveries arising from these works. Some of these beacons from the past seem to contain what we now call UFOs. As with their modern counterparts, the question naturally arises of just what these objects are, and where they came from.

Through Modern Eyes

The images have always been there, in front of our eyes. They may express the real experiences of their makers, or they may simply be anomalous symbols from the hands of those artists depicting their times and the world around them. We have never noticed these images (or is it better to say we have not observed them carefully?) because

Reprinted, with permission, from "Gothic Discs and Renaissaucers," by Daniela Giordano, *Fate*, September 1999.

they are not the main subject of the artwork. They lie in the background, as if the artists were trying to be discrete in communicating their experiences.

A fifteenth-century wood painting on display at the Palazzo Vecchio in Florence, Italy, is one example. The nativity scene *La Madonna e San Giovannino* (*The Madonna and Saint Giovannino*) is often attributed to the Florence painter Filippo Lippi (1406–1469) or his school. In the top right corner of this sacred image, near the head of the Virgin Mary, one finds a curious detail: a lead-gray object with a dome or turret on top. Sloping to the left, the oval object appears to be flying above some kind of barely visible spheroid structure. This mysterious object is characterized by bright yellow-gold rays of light that seem to emanate from its hull.

On the opposite side of the painting is a sun, with three small fires immediately below. These details show that the artist knew the difference between a mystic-symbolic representation and a real event. As if to confirm the artist's will to communicate something of special emotional intensity, a small human figure stands below, observing the object in the sky with his hand shielding his eyes—a sign of attention—while a nearby dog barks at the mysterious object.

Another similar painting is *La Tebaide* by Paolo Uccello (né Paolo di Dono, 1397–1475), kept at the Gallery of Academy in Florence. Uccello was a key figure in developing perspective as a method of representation in paintings. In this piece, he has hidden a dish-shaped object beneath a dome between the umbrella-like sections of some high clusterpines. The ovoid tops of the trees seem to lead into the crucifixion in the background. Uccello illustrates the object's motion with semicircular swirls, as if to indicate a turning—similar to the way printed cartoons represent motion. Moreover, the bright red used by the Aretinian artist to depict the object makes it seem as though he wished to emphasize its brightness.

Shine and Ascend

The ambition to fly has been with humanity since its earliest days, as seen with the myth of Icarus. But the technology leading to airplanes and ultimately spacecraft was not developed until [the twentieth] century—and even then only step-by-step.

A Renaissance painting that hangs above the altar at the church of Saint Peter in Montalcino, Italy, symbolizes a similar evolution of forms. This 1595 painting by Bonaventura Salimbeni represents the Holy Trinity—the Father, the Son, and the Holy Spirit—in accordance with the canons of Roman Catholic tradition. But at the center of the painting, under the large wings of a fading dove that represents the Holy Spirit, sits a round object with two antenna-like stems protruding from it—not unlike the 1950s Russian *Sputnik* satellites that marked the beginning of the space age.

A tapestry by Charles Le Brun (1619–1690)—a painter, sculptor, and early art scholar—similarly hints at today's spacefaring age. One part of the decoration surrounding the main subject of the piece, called *The Four Elements,* consists of a medallion with a picture of something that resembles a modern missile. In addition, the artist has encircled this image with the Latin words *Splendet et Ascendit* (Shine and Ascend).

There are many more artworks whose meanings are unclear, such as the Roman epoch fresco on the Augusto Home wall at Capitolino Hill in Rome. This small untitled painting shows a huge object that resembles a modern rocket ready to take off, as several patricians wearing surprised expressions on their faces watch it from nearby.

A miniature from a French text of 1453 seems to represent some unusual experience that the unknown artist lived through. In the piece, a noble medieval lady wearing a conical hat meets a group of knights, while in the background a mysterious huge gilded sphere with rich decorations hovers over the scene. This miniature might represent an allegorical image, but the detail of a man observing the object with surprise seems to lend it a more literal air.

Perhaps equally baffling is an untitled ancient painting on a wooden furniture drawer from the Earls D'Oltremond in Belgium. It shows Moses receiving the Ten Commandments before what seems to be a squadron of flying craft.

There is a similar fresco on the vault of the Aleksander Nevski Cathedral in Sofia, Bulgaria, completed in 1882. Titled *To Almighty God,* the fresco shows the divine being, head surrounded by bright rays of light, with Jesus in a fairly typical depiction. Just below God's outstretched arm, however, is an object we don't usually see in religious iconography—a round object with a spherical structure at its bottom.

Sense and Sensibilities

It's unlikely these works are all meant as literal depictions of reality, but notice the recurring hints of other beings besides God and his minions inhabiting the sacred realm above. Perhaps they are attempts to understand unusual and inexplicable events witnessed by the artists by placing them within a religious context?

Another miniature from the Renaissance period, *La Contemplazione di San Geremia* (*St. Geremia's Contemplation*), from the *Urbinat Bible* kept in the Vatican Museum, shows that mystic representation, anomalous events, and daily reality are clear in the artist's mind. The town, the mountains and surrounding countryside, the men, and the horses in this piece are perfect expressions of objective representation. The divine image falls in the classic patterns of religious iconography, thus an object in the sky seems to represent an unusual visual experience.

A sphere floating in the upper right corner of the painting emits blazing rays of curving light above the divine images. A straight beam of light seems to come from within the object. Does it perhaps represent a meteoric fireball with an unusual tail? Maybe, but we can never know for certain what the artist saw. One thing is clear, as with all of these puzzling works: He wanted to tell us something.

Is it possible the unusual objects in these historic artworks are nothing more than allegoric images stemming from the inspiration of those who painted them? If so, what are they meant to symbolize?

It seems that these paintings depict aircraft that simply can't be explained by the technology of those centuries. Thus we may entertain the possibility that they are UFOs. Without knowing just what the artists saw, however, we are left to wonder what these objects are and where they came from—disquieting questions with no definitive answers.

UFOs in Popular Culture

Phil Patton

In the following essay, Phil Patton examines the history of flying saucers, the first of which was reported more than fifty years ago, and the rapid adoption of UFOs into American mythology. Flying saucers have become part of American popular culture, Patton writes, and appear regularly on television and in movies, advertising, and art. He discusses how the myth of flying saucers has grown and changed over the years, influencing American culture in the process. Patton is the author of *Dreamland: Travels Inside the Secret World of Roswell and Area 51.*

While flying near Mount Rainier in Washington on the afternoon of June 24, 1947, Kenneth Arnold spotted nine shiny, bright objects in loose formation in the distant sky. They were shaped like boomerangs or flying wedges and moving at tremendous speed.

After landing in Pendleton, Ore., Arnold, a veteran pilot and successful businessman, described his sighting to Nolan Skiff, a columnist for the *East Oregonian,* the local newspaper, saying the objects "flew like a saucer would if you skipped it across the water." Skiff's account was soon picked up and carried across the country by the Associated Press. In the process, Arnold's description of the objects "flying like saucers" was transformed into "saucerlike" objects and ultimately "flying saucers."

It has been half a century now since Arnold's sighting, and the flying saucer still zips across the skies of popular culture with dependable regularity. Films from 1996's *Independence Day* and *Mars Attacks!* to *Men in Black* offer fresh views of the flying disks. They crop up in television shows like *The X-Files* and *Dark Skies.* They hover in malls, emblazoned on T-shirts and CD covers. They also inspire painters, architects and designers.

Modern Mythology

The image of the flying saucer is at the center of a modern mythology, a figure of folklore, focusing fears and hopes like the lens whose shape it shares, reflecting the wider culture in its mirrored surface.

Nothing says more about the origins of the flying saucer myth than the birth of the name in the press. The flying saucer became a new kind of mythological figure—akin to a Hermes or a Puck, a unicorn or a leprechaun—that flourished primarily not in oral or literary tradition but in the mass media. The first folk figure to emerge from the realm of technology, it is the most flexible of cultural icons, with overtones ranging from the cosmic (dark visions of space invaders) to the comic (cartoons inhabited by stubby flying saucers piloted by little green men).

Arnold's sighting was at first treated as a novelty, but within days reports of dozens of others flowed in from around the world. In July, the Air Force issued a press release claiming that a flying disk had been "captured" in Roswell, N.M.; it later decided that the object had been a weather balloon. (The Roswell incident was quickly forgotten, only to be reexamined in the 1980's in books and television shows.) Within two months, polls showed that 90 percent of Americans had heard of flying saucers.

America in 1947 was in the midst of vast technological change: the transistor was introduced, the sound barrier broken, the first earth satellites were proposed. Politically, the battle lines for the cold war were being drawn, and with the advent of the atom bomb, the threat of worldwide Armageddon seemed palpable. The historian Curtis Peebles, author of the 1994 book *Watch the Skies! A Chronicle of the Flying Saucer Myth,* reflecting on the time, asks, "Why not flying saucers?"

By the end of the 40's the Air Force had introduced the less judgmental term "unidentified flying object," but flying saucer as a term and an image was already well established in the popular consciousness.

One of the first to present the flying-saucer phenomenon as mythology or folklore was the psychoanalyst Carl Jung. He had paid attention to even the earliest reports about them, and in 1957 he published *Flying Saucers: A Modern Myth of Things Seen in the Skies.* In it, Jung made no attempt to answer the question of whether they existed or not. For him, the flying saucer was an archetype in the making, an icon, a modern mandala, embedded deep in the collective unconscious. "Such an object," he wrote, "provokes, like nothing else, conscious and unconscious fantasies."

"Our time is characterized by fragmentation, confusion and perplexity," he added. "At such times men's eyes turn to heaven for help, and marvelous signs appear from on high."

Almost from the beginning, flying saucers were associated with a familiar set of narratives that were folkloric in nature. Some were hopeful (the warning messages from distant planets), others sinister (a Government cover-up of captured saucers). Those who claimed space travelers had contacted them, like George Adamski and George Van Tassel, inspired near-religious followings. By 1954, conventions at Giant Rock in the Mojave Desert were drawing up to 5,000 people. At roughly the same time, Maj. Donald Keyhoe, author of the best seller *The Flying*

Saucers Are Real, was accusing the Government of hiding the truth.

Many of the fantasies Jung described were made manifest in movies. In *The Day the Earth Stood Still* in 1951, the occupants of the flying saucers brought stern but beneficent warnings about the dangers of atomic bombs. In 1956, *Earth Versus the Flying Saucers* warned that the launching of satellites might provoke an alien attack. In the 1953 movie *Invaders From Mars,* spacemen were surrogates for communism; Mars, after all, is the Red Planet.

Flying Saucers in Modern Art

But the flying saucer myth also seemed to grow from the earlier, more profound sense of dream and dislocation of Dada and Surrealist art. In the 1920's, Meret Oppenheim had covered a saucer—along with a teacup and spoon—with fur to produce one of the classic Dadaist objects. That work seemed to emerge from the same sensibility that produced Man Ray's giant lips in the sky and Magritte's suspended bowler hats.

Another old theme on which the flying saucer narrative drew was that of the mysterious stranger. Simply believing in saucers made one socially suspect. Moreover, it was among those who felt culturally alienated that flying saucers had their greatest appeal. Flying saucers came to figure heavily in the work of outsider artists like Alexander Maldonando and the Rev. Howard Finster. The artist Ionel Talpazan views his brightly colored drawings of flying saucers as technical diagrams that explain their propulsion systems. . . .

Mr. Talpazan's life is a study in alienation. Born in Romania during the oppressive rule of Nicolae Ceausescu, Mr. Talpazan was abused as a child; he also claims to have seen his first flying saucer at age 8. He came to the United States, where he was inspired to draw flying saucers by a television documentary on U.F.O.'s. He now sees himself as a kind of unpaid consultant to NASA, but what his alien ships reflect most may be his own sense of alienation.

A Transformation

In the half century since Arnold's sighting, flying saucer lore has grown and changed. In the 50's, the shape of the flying saucer seemed to reflect the optimistic curve of rising prosperity and unfolding technology. It also evoked the orbits of electrons around the nucleus of an atom or those of satellites around the earth. Artists and design artists were inspired by the shape. It cropped up in the design of fast-food outlets and airport terminals as well as the Space Needle in Seattle and John Laudner's Chemosphere house, which hangs over the Hollywood Hills.

But the image of the flying saucer would sail too across the animated frames of the television show *The Jetsons* and, beginning in 1957, across American playgrounds in the form of the Frisbee. One early model of the plastic toy disk, marketed by the Wham-O Company, was called the Flying Saucer. The next year, Billy Lee Riley and his Little Green Men had a rockabilly hit with "Flying Saucer Rock-and-Roll." A similar cartoon playfulness informed Keith Haring's saucer drawings of the 1980's.

The flying saucer has shown an amazing capacity for reinvention, yielding new messages for new generations. So *Independence Day* can shamelessly reprise *The War of the Worlds*, substituting a computer virus for the bacteria of the original H.G. Wells story. And *Mars Attacks!* offers a campy take of images from a 1960's set of flying saucer trading cards. In the 1950's, people reported being taken aboard saucers by "happy space friends"; in the 1980's, abductees reported darker, more manipulative experiences. The imagined occupants of the saucers have changed, too, from the kindly extraterrestrials of Steven Spielberg to the chortling, evil imps of *Mars Attacks!* "Aliens," says Barry Sonnenfeld, the director of *Men in Black*, "make perfect bad guys. You have no problems of political correctness."

The makers of *Men in Black* give their own twist to an old story. Tommy Lee Jones and Will Smith play agents of a kind of intergalactic immigration agency, as the movie ads proclaim, "protecting earth from the scum of the universe." The film is based on a comic book by Lowell Cunningham, who says he heard the story from a friend.

Like the flying saucer myth, the man-in-black story began with a specific incident. A few days after Arnold's sighting, Fred Lee Crisman and Harold A. Dahl reported gathering wreckage from a saucer that crashed on Maury Island, Wash. They said a mysterious man in a black suit had then appeared, warning them to keep quiet. Their story proved to be a hoax, but retold in a book, *They Knew Too Much About Flying Saucers* by Gray Barker, the man-in-black myth became a part of the saucer myth.

The transformation of the story from first a press report to a folkloric tale to a comic book and now to a film illustrates how the myth is transformed. That process is not unlike the children's game of telephone or what the literary critic Harold Bloom calls innovation by misinterpretation.

The period when the saucer phenomenon was new is now a point of reference. For *Men in Black*, the production designer Bo Welsh consciously sought out settings that evoke what Mr. Sonnenfeld calls flying saucer architecture: the Trans World Airlines terminal at Kennedy International Airport, the Unisphere, the Guggenheim Museum. Mr. Sonnenfeld says that he tried to make the film walk the line between dark comedy and campy parody trod by Stanley Kubrick in his 1964 film *Dr. Strangelove*.

After a half century of mythologizing, today's representations of flying saucers glint with a self-conscious wit. Those crescent shapes that Arnold saw in the sky near Mount Rainier are starting to look more like something else: quotation marks.

UFOS IN THE MOVIES

Paul Meehan

Paul Meehan is the author of *Saucer Movies: A UFOlogical History of the Cinema*, from which the following selection is excerpted. Meehan writes that "saucer movies"—a subgenre of science fiction films that depict the first contact between aliens and humans—have been a part of American popular culture since the beginning of the motion picture era. According to Meehan, saucer movies have moved past mere entertainment and are now a sociological phenomenon. He presents a number of hypotheses to explain the connection between the movies and the phenomena of abduction stories, UFO sightings, and conspiracy theories.

Flying saucers and motion pictures have been involved in a love affair for a hundred years. Long before the advent of the modern UFO era, strange craft were flying through celluloid skies and alien beings were visiting the Earth from other worlds in darkened theaters.

Beginning with Kenneth Arnold's classic sighting of "flying saucers" in the summer of 1947, the phenomenon of Unidentified Flying Objects has become part of our popular culture, especially in American films from the 1950s to the present. Some films on the extraterrestrial theme, notably *E.T.*, *Independence Day* and the *Star Wars* series, are among the most popular movies ever made. Many of Hollywood's most respected directors, from Howard Hawks and Robert Wise to Steven Spielberg and Ridley Scott, have contributed to the genre.

UFOs are inherently cinematic, offering a filmmaker nearly infinite possibilities, especially as film technology in the 1990s approaches the level of science fiction. Special effects teams using the latest in computer technology and other exotic hardware can conjure extremely convincing visions of alien glory in 70mm, THX digital sound. The UFO theme lends itself to any number of genres, including: horror (the *Alien* series); comedy (*Visit to a Small Planet, Spaced Invaders, Coneheads, Men in Black*); drama (*Starman, Batteries Not Included, Cocoon, Phenomenon*); and racial allegory (*The Brother from Another Planet, Alien Nation*).

Saucer Movies

Saucer movies are a distinct subgenre of the science fiction film, and perhaps should constitute a genre of their own. They are separate and distinct from a related subgenre, that of the Space Opera, best typified by the popular *Star Wars* and *Star Trek* series. In Space Opera, contact between the human race and alien cultures is a *fait accompli*, yet humans are still the dominant life form. Aliens depicted in these films are usually exotic, second-banana characters, like the profusion of weird beings in the cantina sequence of *Star Wars*. Space Opera almost never takes place in the present or on the planet Earth, but in the future or in long-ago galaxies far away, when human technology has achieved interstellar space flight and mankind has encountered alien races among the stars, on more or less equal terms. War and adventure are the dominant motifs. Space Opera had its genesis in the *Buck Rogers* and *Flash Gordon* comic strips of the 1930s. It has proven to be a popular form right up to the present.

Saucer movies are all about the first contact between humankind and alien civilizations. They are usually, with a few significant exceptions, set in the present, on Earth. Aliens are always represented as visitors or invaders, possessing a superior technology that threatens or challenges mankind. Saucer movies evolved in the early 1950s, not from literary science fiction sources that had been around for decades, but as a response to UFO sightings in the early post–World War II era. They attempt to deal with the philosophical implications of the first encounter between human beings and alien cultures. This first contact can be shown in many ways. Sometimes only an alien artifact is found, as in *2001: A Space Odyssey* or *Forbidden Planet*. Communication can be made via radio (as today's Search for Extraterrestrial Intelligence, or SETI program proposes) as in *Contact*. Most of the time, however, the aliens arrive in wondrous celestial ships, bearing messages of hope or doom. There is almost always a feeling of dread associated with this contact. No matter how benevolent the visitors from beyond turn out to be, there is always an uneasy mood of xenophobia that lies just beneath the emotional surface of these films.

Saucer Movies and the UFO Phenomenon

Unlike other types of science fiction films, saucer movies interface with a sociological event, the UFO phenomenon. This phenomenon has its own rich mythology, which has evolved from early tales of distant nocturnal lights and daylight discs to frightening stories of landings and abductions. Many of the UFO reports of the 1990s seem like sci-fi movie plots of the 1950s. Current UFO research centers around two issues: abductions and cover-ups. Abduction narratives relate to how humans have been taken by aliens inside UFOs and subjected to various physical and psychological ordeals in connection with a genetic manipulation program. Cover-up stories tell of captured saucers, recovered alien corpses, CIA involvement and secret deals with the UFO-nauts. Both of these themes have a paranoid flavor suggestive of classic black and white invasion films of yesteryear.

Saucer movies are probably the major shaper of the UFO mythos, along with supermarket tabloids and television shows. Between these films and reported UFO events there has been a cross-pollination of ideas. Specific details created in saucer movies are replicated in alleged UFO events, while incidents from UFO encounters find their way into science fiction films. Each side functions as a mirror image of the other. To what extent does popular culture overlay our perceptions and attitudes about the UFO phenomenon?

Most researchers tend to disassociate UFOs from the influence of the movies. They imply that the perceived reality of UFOs may be compromised by being considered as a product of the Hollywood dream factory. UFOs would then be reduced to a figment of cinematic imagination, an optical illusion that somehow escaped into reality. Researchers have pointed out that there was no upsurge in UFO sightings after the release of Steven Spielberg's *Close Encounters of the Third Kind* in 1977 or *E.T.* (now the third most popular film ever made) in 1982. Considering the enormous popularity of the *Star Trek* films and TV shows, the *Star Wars* series and *E.T.*, aliens fitting the description of the pointy-eared Mr. Spock, the gnomelike E.T. or the exotic creatures from *Star Wars* have not been sighted aboard UFOs. Reports of alien abductions of humans cannot be ascribed to the influence of film, as there have only been four movies depicting the classic "abduction scenario," namely *The UFO Incident* (1975), *Communion* (1989), *Intruders* (1992) and *Fire in the Sky* (1993). Of these, only *Communion* and *Fire in the Sky* had a regular theatrical release. *The UFO Incident* and *Intruders* were produced for television, reaching a wider audience but on a smaller screen.

During the formative period of the 1950s and 1960s, the dwarfish humanoids referred to as "Grays," which are so prevalent in today's UFO literature, were conspicuously absent from the screen. For budgetary reasons, most aliens bore a remarkable resemblance to normal human beings. There were distinguished gentlemen from outer space, such as Michael Rennie's classy Klaatu in *The Day the Earth Stood Still* (1951), Helmut Dantine's lofty Venerian in *Stranger from Venus* (1954), or John Carradine as *The Cosmic Man* (1959). These godlike visitors no doubt inspired the "contactees," individuals like George Adamski, Daniel Fry and Truman Bethurum, who claimed to have been contacted by good-looking, humanlike aliens bearing messages of world peace. Another approach was to construct a nasty alien out of an oversized actor in heavy make-up, such as James Arness in the original version of *The Thing* (1951). Many invaders were simply men dressed in tight fitting uniforms with hoods, notably Leonard Nimoy in *Zombies of the Stratosphere* (1952). It was much riskier to depict more exotic alien creatures, as this frequently descended to the ludicrous. The "gumdrop" monster of *It Conquered the World* (1956) and the crawling carpets of *The Creeping Terror* (1964) are two of the more amusing examples. A few movies managed to summon the imagination and special effects expertise to convincingly render non-humanoid aliens, such as the cyclopean "xenomorphs" of *It Came from Outer Space* (1953), the gelatinous menace of *The Blob* (1958), and the trilobite Martians of *War of the Worlds* (1953), but these films were the exceptions rather than the rules. Some thought may have been given to avoiding the cliche of making the aliens look like "little green men." Only two films of the period, *The Man from Planet X* (1951) and *Invasion of the Saucer Men* (1957), went this route by depicting aliens as small humanoids with outsized heads. . . .

The year 1966 saw the publication of John Fuller's *The Interrupted Journey*, an account of the abduction experience of a New England couple named Betty and Barney Hill. They claimed that they were abducted from their automobile late at night by small aliens with gray skin, large heads and large eyes, taken aboard a saucer and given a series of medical examinations. The Hills would emerge as the first popularly known "abductees." . . . The abductees told frightening stories of alien abductions at the hands of the enigmatic, inhuman Grays.

UFO skeptic Philip J. Klass has suggested that the first TV broadcast of *The UFO Incident* (based on the Betty and Barney Hill abduction story) on October 20, 1975, provided thematic material for a rash of subsequent abduction reports, the most famous being that of Travis Walton, reported about two weeks later on November 5, 1975. Walton's abduction story was later filmed as *Fire in the Sky* (1993). Klass argues that several other abduction accounts reported in the UFO literature in 1975 . . . first came to light around the time of the broadcast. He therefore contends that this single TV movie-of-the-week led to a mushrooming of abduction reports generated by fantasists and liars out for notoriety and personal gain. . . .

The Bad Film Hypothesis

Another argument, which could be called "Bad Film Hypothesis," has been put forth by researcher Martin Kottmeyer. He points to the fact that many themes from UFO abduction lore had previously appeared in a number of forgettable, grade-Z flicks from the 50s. . . .

According to the Bad Film Hypothesis, our subconscious minds have been fed a steady diet of sci-fi mythology that has been forgotten and then mentally regurgitated in the form of UFO abduction tales. Much of this conditioning presumably occurred with the subject in a hypnotically suggestible state, that is, during late night TV viewing. Many correspondences between UFO lore and half-remembered bad films can be cited. Aliens have been abducting humans on the screen since *The Man from Planet X* arrived in 1951. Kottmeyer claims to have found a correspondence between Barney Hill's description of extraterrestrials with wraparound, telepathic eyes and an episode of the sci-fi TV series *The Outer Limits* entitled "The Bellero Shield." The episode, which aired shortly before Barney Hill made his description, featured

an alien with wraparound, telepathic eyes. In fact, these two characteristics have been described in many abduction cases since the Hills'. In
this way, fictional information from the mass culture is transformed
into nonfiction UFO reports, although this theory assumes that Barney
Hill must have actually watched "The Bellero Shield." Kottmeyer further argues that UFO investigators systematically remove the more
obvious cultural references in their reports. Therefore, the only cultural data that are going to sneak by are from obscure sources like *The Outer Limits* and *Killers from Space*. He points out that the bad film buff has
an advantage in knowing about obscure cultural material that may
find its way into descriptions of "real-life" UFO events.

Kottmeyer suggests that it took about a decade worth of "psy-
chotronic" saucer movies incubating on the back burners of the mass
consciousness to start producing abduction reports with similar
themes. Yet there were only a handful of cases reported during the 60s,
after ten years of bad film indoctrination. The mass of abduction
reports began surfacing in the mid 70s, and became yet more numerous through the 80s and early 90s. At the same time, due to the advent
of color television, the creaky black and white creature features of the
50s had been phased out of broadcast schedules during this period. In
order to view most of these films today, you must purchase them
through mail-order companies that specialize in bad films, as most
video stores do not stock them.

The Prescience Hypothesis

There may be another way of looking at the connection between
saucer movies and abduction themes. If the assumption is made that
the mass of abduction stories are factual, it is possible that film provides a kind of mirror image of the reality of the UFO phenomenon,
transmitted through the focusing lens of the imagination. Our first
contact with a nonhuman civilization would be the greatest event in
human history. In the past, art has been observed to manifest a degree
of prescience. Real-life events such as the sinking of the *Titanic* and the
kidnapping of Patty Hearst had been written up in novels years before
they actually occurred. Movies like *Suddenly* (1954) and *The Manchuri-
an Candidate* (1962) seem to have anticipated the Kennedy assassina-
tion. The most dramatic example in recent memory involved the
nuclear melt-down depicted in *The China Syndrome* (1979), released pri-
or to the atomic accident at Three Mile Island nuclear power plant in
Pennsylvania. There was even a chilling line of dialogue in the film to
the effect that if the reactor exploded it would devastate an area "the
size of Pennsylvania." Similarly, Robert Wise's film *The Day the Earth
Stood Still,* released in September of 1951, featured a flying saucer landing in Washington, D.C. This seemed to prefigure the spectacular
radar-visual UFO sightings over Washington that panicked the nation
in July 1952.

More recent examples of prescience in film would include Christopher Reeve's role in a thriller entitled *Above Suspicion* (1995), in which he played a wheelchair-bound cop paralyzed in the line of duty. This was his last movie role before being paralyzed in a tragic riding accident and confined to a wheelchair himself. *The Trigger Effect,* a film about social chaos caused by a widespread power blackout, was released in the summer of 1996, a few weeks before a massive electrical failure blacked out cities across the western United States. More recently, the film *Wag the Dog* anticipated the crises in the Clinton presidency in early 1998. Undoubtedly the most eerie saucer movie coincidence involved the 1979 TV-movie *Mysterious Two.* Inspired by a bizarre UFO cult led by a couple named "Bo" and "Peep," *Mysterious Two* follows the strange pair as they lead their flock away into the wilderness toward an uncertain fate. On March 26, 1997, police in San Diego, California, discovered the bodies of thirty-nine members of the Heaven's Gate cult, whose leader, Marshall Herff Applewhite, was the cult leader depicted in *Mysterious Two.*

Crossbreeding

Certain stock themes from saucer movies seem to manifest a change in meaning in light of material that has emerged quite recently from the abduction material collected by Budd Hopkins, David Jacobs and others. According to these reports, the aliens are heavily involved in a program of genetic manipulation, hybridization or apparent crossbreeding between their species and ours. Women have been artificially impregnated, and later have had the fetuses removed.

An emphasis on human/alien crossbreeding has been a consistent theme of UFO films since *When the Man in the Moon Seeks a Wife* (1908). In the 50s, when saucer movies aimed at the titillation of the youth market, this theme became popular in films such as *I Married a Monster from Outer Space* (1958), *Devil Girl from Mars* (1954), *Village of the Damned* (1960) and *The Mysterians* (1959). The most haunting example, however, is from the concluding scene of Stanley Kubrick's *2001: A Space Odyssey* (1968). Astronaut Frank Bowman (Keir Dullea) is abducted by an alien force and transformed into a gigantic, god-like fetus floating in space, regarding the Earth with enormous, allknowing eyes.

This mystifying image was interpreted through the filter of the psychedelic 60s subculture at the time of the film's release. The giant fetus looming over the Earth was thought to represent the "Star Child" dawning on the Age of Aquarius. Seen in the light of the abduction material of the 80s and 90s, this same image assumes an entirely different significance, one that Kubrick did not consciously intend. Although the Betty and Barney Hill case, including Betty Hill's needle-in-the-navel "pregnancy test," had been written up in the early 60s, there were no stories of alien breeding programs at the time *2001* was made to suggest a relationship between alien intelligences and human fetuses. . . .

Crash and Retrieval Stories

Besides abduction reports, the other burning question in UFO research in the 80s and 90s has centered around the investigation of "crash-retrieval" cases, tales involving the acquisition of downed saucers by the U.S. armed forces and subsequent cover-up by the military, the CIA, and various ultrasecret alphabet soup agencies. The most famous of these is usually referred to as "The Roswell Incident," concerning the retrieval of a crashed saucer and its dead occupants near Roswell, New Mexico, by the Air Force in 1947. It has been alleged that the U.S. government has established a super-secret project, called MJ-12, to study the aliens and captured UFO technology. In order to prevent panic, a massive cover-up has been instituted by the government to keep the facts from the public. It has even been suggested that covert contact has already taken place, according to some accounts, as long ago as during the Eisenhower administration. Crashed saucers are reportedly held in a mysterious facility known as "Hangar 18," usually located in Wright-Patterson Air Base in Dayton, Ohio. The government is supposedly testing alien technology at a super hush-hush facility called "Area 51," somewhere near Groom Lake, Nevada.

The UFO cover-up stories have led to some curious inferences about saucer movies. Researchers have suggested that Hollywood films are being used by the government as a means to prepare the general public for the shocking truth that UFOs exist. These allegations have generally been leveled at the work of Steven Spielberg, whose films *E.T.* and *Close Encounters* were enormously popular. Curiously, both of these films had subplots involving secret government agencies covertly studying UFOs. There is an apocryphal story about Spielberg and former president Ronald Reagan, which reportedly occurred during a private screening of *E.T.* at the White House in the summer of 1982. Reagan is said to have whispered to Spielberg that "there are probably only six people in this room who know how true this is."

Researcher and film producer Linda Moulton Howe has recently alleged that, in connection with research she was doing for a UFO documentary in 1983, government sources informed her that Robert Wise's 1951 first contact classic *The Day the Earth Stood Still* was, in her words, "one of the first government tests of public reaction to such an event." She suggests that producers, directors and screen writers have been working behind the scenes to plant the idea of UFOs into the public mind as far back as the 1950s. This amounts to using the existence of UFO films as evidence for the reality of the UFO phenomenon.

Crash-retrieval themes in saucer movies go back to the very first film ever made about the subject in the post–Kenneth Arnold period, *The Flying Saucer* (1950). In this film a U.S. secret agent (played by Mikel Conrad, who also produced and directed) is sent to Alaska to find a downed saucer. He eventually discovers the craft, along with a group of Soviet scientists, but the flying saucer turns out to be nothing more than an advanced, experimental aircraft being flown by an American pilot (Denver Pyle) trying to keep it away from the prying eyes of the Commies.

Another early crash-retrieval treatment was *The Thing* (1951), in which a saucer and its single occupant are discovered frozen under the Arctic ice by members of a military-scientific expedition near the attempted recovery, but the occupant is hacked out of the ice and taken back for study. Although based on John W. Campbell's 1938 classic short story "Who Goes There?", elements of the plot seem to echo accounts of the Roswell Incident, with the locale changed from the desert to the Arctic. . . .

A Metamorphosis

Saucer movies have gone through a metamorphosis over the years in terms of the treatment of the UFO theme. In the paranoid 50s, the coming of the saucers seemed to herald the apocalypse. The vast majority of these early UFOnauts were hostile monsters bent on the enslavement or destruction of the human race. This xenophobia is still prevalent in films today. In the late 60s, however, the image of the aliens began to soften a bit. Kubrick's *2001* depicted the first human-alien encounter in a poetic and mystical light, as a rebirth of the human spirit. At the same time the first abduction reports became manifest in the public mind. In these early reports the aliens seemed like curious scientists who were merely studying the Earth, just exobiologists like little E.T. This was reflected in films of the 70s and 80s, not only in the Spielberg productions *Close Encounters* and *E.T.,* but also in movies like *Starman* (1984) and *Cocoon* (1985). In these films the aliens are givers of gifts that will transform our society.

In the late 80s other twists on the theme began to appear. Saucer movies got funny. Comic variations on the theme came into vogue in genre entries like *Martians Go Home* (1990), *Killer Klowns from Outer Space* (1988), *Earth Girls Are Easy* (1989), and *My Stepmother Is an Alien* (1988). Even Woody Allen got into the act. In *Stardust Memories* (1980), Woody has a close encounter with a group of little aliens who urge him to "tell funnier jokes." There were also a number of films that moved the saucer movie into the action-adventure and police thriller genres, such as *The Hidden* (1987), *Predator* (1987), *Alien Nation* (1988), *Abraxas* (1990), *I Come in Peace* (1990) and *Aliens* (1986).

After the publication of updated abduction narratives such as Budd Hopkins' *Intruders,* Whitley Strieber's *Communion* and David Jacob's *Secret Life* in the late 80s–early 90s, saucer movies began to draw their inspiration from these bizarre tales. *Communion* and *Intruders* were filmed. The Travis Walton abduction case was given a highly dubious screen treatment in *Fire in the Sky.* These films, like the real-life events that inspired them, show the ETs in an ambiguous light. They must attempt a quasi-documentary treatment of events "based on a true story," yet are highly fictionalized. *Communion* (1989), which was copro-

duced by fiction writer Whitley Strieber, author of the nonfiction best-
seller, is an idiosyncratic and rambling work that sometimes descends
into the ludicrous. *Intruders* (1992) was inspired by Budd Hopkins'
account of the abduction experiences of a woman he called "Kathie
Davis." The TV miniseries, cowritten by UFO researcher Tracy Torme
(son of songster Mel Torme), instead presented a highly fictionalized,
generic abduction narrative incorporating little of Davis' story. The
1975 Travis Walton abduction was filmed as *Fire in the Sky* (1993). Also
cowritten by Torme, the script deals with the human interrelationships
in the Walton Case, yet totally distorts Walton's abduction account in
favor of a grotesque horror treatment.

Of these three films, only *Intruders* treats the subject intelligently.
Communion and *Fire in the Sky* opted for cheap horror movie thrills, and
actually spread misinformation about this complex phenomenon. By
way of contrast, both *The UFO Incident* and *Intruders* portrayed the
abduction phenomenon open-mindedly and accurately according to
the published UFO literature, perhaps because they were produced for
TV and not for theatrical release. True-life UFO abduction movies have
never done well at the box office, and probably never will.

Full Circle

By the 90s the abduction scenario had all but vanished from movie
screens, replaced by paranoid visions of conspiracies and cover-ups
inspired by the success of Fox-TV's popular show *The X-Files* and in
films like *Roswell* (1994) and *Men in Black* (1997). Saucer movies
became more popular than ever in blockbusters like *Stargate* (1994),
Species (1995) and *Independence Day* (1996). The 50s-style alien invasion themes came back into vogue in films like *The Arrival* (1996), *Mars
Attacks!* (1996) and *Starship Troopers* (1997). The genre seems to have
come full circle, returning to the invasion-oriented themes so prevalent in the 50s.

The overall relationship between UFOs and saucer movies is complex and multifaceted. The Bad Film Hypothesis, the Government
Conspiracy Hypothesis and the Prescience Hypothesis are all ongoing
debates, removing the subject of saucer movies beyond simple film
criticism and history, into the realms of sociology, political science and
even paranormal research.

French researcher Jacques Vallee uses cinema as a metaphor for the
UFO phenomenon itself. According to Vallee, UFOs, like the cinema,
constitute a "meta-system" for disseminating information. Instead of
concentrating on the saucer movie light show, Vallee yearns to find
the projection room to examine the deeper process behind the
image. Perhaps the study of saucer movies can shed some light on
the ultimate intelligence, human or otherwise, behind the UFO phenomenon.

UFOs AND RELIGION

Irving Hexham and Karla Poewe

Unidentified flying objects have been closely associated with religion and spirituality practically since the first flying saucer craze of the 1950s, report Irving Hexham and Karla Poewe. The authors examine several religions and cults that have developed out of the fascination with UFOs, including Scientology. They also discuss the cult of Heaven's Gate and the 1997 incident in which thirty-nine of its members committed suicide so that they could be "transported to a higher level" via a UFO that they believed was flying behind a comet. Hexham and Poewe, professors of religious studies and anthropology at the University of Calgary, are the authors of *New Religions as Global Cultures: Making the Human Sacred.*

Years ago British anthropologist E.E. Evans-Pritchard went to live among the Azande people in Africa. An educated person of the West and a social scientist, Evans-Pritchard rejected the witchcraft beliefs of the Azande. But he began to recognize that to live and work among the Azande one had to assume the reality of witchcraft. Once this was done, social practices fell into place and the world made sense. He found that to live in an Azande village required a leap of the imagination without which it was impossible to obtain the basic necessities of life. Only by acknowledging the reality of witchcraft could he negotiate the basic transactions needed to keep him alive.

Attempting to understand the logic that led 39 well-educated people to commit suicide in San Diego in 1997 because they believed they would be transported to a higher level of being by way of a spacecraft that was tailing the Hale-Bopp comet requires a similar leap of the imagination. Usually we associate UFOs with science, possible other worlds and hard-nosed science fiction. Most discussions of UFOs concern 1) the question of whether there is scientific evidence for them and 2) theories about extraterrestrial life. All this seems a far cry from the pseudo-theosophical and gnostic ideas propagated on the Heaven's Gate Higher Source Web site.

But at least since the time of the first "flying saucer" craze of the early 1950s, interest in UFOs has been closely tied to matters of spirituality.

Reprinted, with permission, from "UFO Religion," by Irving Hexham and Karla Poewe, *The Christian Century*, May 7, 1997. Copyright © 1997 Christian Century Foundation.

The suggestion that extraterrestrials regularly visit the earth was first made by the American journalist Charles Fort (1874–1932). In *The Book of the Damned* (1919) and other books, Fort argued that modern science represented a new kind of "priestcraft," which, he claimed, refused to admit certain inconvenient truths. Presenting a monistic vision of the universe, Fort systematically replaced Christian ideas of creation and providence with a form of secular spirituality involving godlike extraterrestrials.

Many early science fiction writers, including Damon Knight (1922–), Eric Frank Russell (1905–1978) and Sam Moskowitz (1920–), were influenced by Fort. More important, his ideas about extraterrestrials observing and guiding human development were adapted by E.E. "Doc" Smith (1890–1965) in *Triplanetary* (1934), which he developed into his "Lensmen" series in 1948. This highly popular series united numerous spiritual ideas and mythological themes into a hi-tech space opera. George Lucas's *Star Wars* movies were consciously modeled on the Lensmen books.

Another writer influenced by Fort was Richard Shaver (1907–1975), who created a sensation in March 1945 when his story "I Remember Lemuria" appeared in *Amazing Stories*. This fantastic yarn about lost civilizations generated an intense controversy. A series of sequels followed, leading to the publication of the books *I Remember Lemuria* and *The Return of Santhanas* in 1948.

The first UFO books appeared in 1950. Most of these were uninteresting descriptions of strange lights in the sky. In 1953 Desmond Leslie and George Adamski published *Flying Saucers Have Landed*. Adamski claimed to be the first human to have encountered space aliens visiting earth in UFOs. Significantly, both Adamski and Leslie, like Fort and Shaver before them, engaged in theosophical speculation. Long before they used science fiction to transform theosophical concepts into pseudoscientific claims about UFOs, these writers were deeply immersed in occult literature. From the beginning, UFO stories were entangled with religious beliefs of theosophical origin supported by rich occult mythologies.

Following the success of Leslie and Adamski, a host of other spiritually inclined writers made similar claims. The most important of these was Erich von Daniken, whose book *Chariots of the Gods?* (1968) purported to be a serious study of strange evidence suggesting that spacemen once visited the earth. Shaver's influence on von Daniken is clear. Later, in *The Gold of the Gods* (1972), von Daniken dropped his pseudoscientific stance to reveal his true religious interests. Other books, like Brad Steiger's *Gods of Aquarius* (1976) and Jacques Vallee's *Messengers of Deception* (1979), continue to blur the distinction between science and religion, empirical reality and the occult.

Surveys show that between 70 and 75 percent of North Americans believe in extraterrestrials and UFOs. Consider also the much smaller, but growing, number of people who claim contact with UFOs and/or abduction by space aliens, and the plot thickens.

Scientology

The best-known religion that has developed out of a fascination with UFOs is L. Ron Hubbard's Scientology. Hubbard's first works were science-fiction adventure stories. In 1938 he published "The Dangerous Dimension" in the magazine *Astounding*, and he eventually became one of the magazine's most prolific writers.

In 1950 he published *Dianetics*, which he proclaimed as a "new science of the mind." In the background of this science was Hubbard's fascination with interplanetary travel. One of the earliest and most enthusiastic converts to Dianetics was none other than *Astounding*'s charismatic editor John W. Campbell, who did all he could to promote Hubbard's views through his magazine. Other science fiction writers such as Katherine MacLean, James Blish and Kurt van Vogt were drawn into the enthusiasm, although later all moved away from Hubbard's movement.

Various other new religions were founded by people who had dabbled in Scientology, the most successful being EST and Eckankar. The most infamous is the Heaven's Gate community, whose leaders studied Scientology in the early 1970s.

Other UFO Cults

Meanwhile, other UFO cults emerged. These included the Wallace Halsey's Christ Brotherhood; the Association of Sananda and Sanat Kumara made famous in *When Prophecy Fails* (1955), written by Leon Festinger, Henry W. Riecken and Stanley Schachter; and George King's Aetherius Society, founded in 1955. Today the Unarius movement, which expects salvation in 2001, and the Realians, who say that 2035 is a more likely date, are the most influential of these movements. More recently, UFO beliefs have gained ground among fundamentalist Christians through the writings of men like Texe Marrs and Gary North. They give UFOs a demonic spin, regarding them as part of a cosmic conspiracy threatening Christianity.

Behind all these movements and beliefs lies a mythology of creation which rejects evolution as a scientific concept but which, except in the case of the fundamentalists, cannot accept a biblical view of creation. UFOs are regarded as the vehicles of creation, providence and final salvation. This spiritualized universe resembles early gnosticism in its emphasis on escaping from earthly existence into lost worlds and other civilizations that provide a higher (nonmaterial) realm of existence. Modern gnostics would object to applying the term to Heaven's Gate, but the religious themes are indeed strikingly similar.

ORGANIZATIONS TO CONTACT

The editors have compiled the following list of organizations concerned with the issues presented in this book. The descriptions are derived from materials provided by the organizations. All have publications or information available for interested readers. The list was compiled on the date of publication of the present volume; the information provided here may change. Be aware that many organizations take several weeks or longer to respond to inquiries, so allow as much time as possible.

Center for the Study of Extraterrestrial Intelligence (CSETI)
PO Box 265, Crozet, VA 22932-0265
(301) 249-3915
website: www.cseti.org

A research and educational organization, CSETI is dedicated to establishing peaceful and sustainable contact with extraterrestrial life-forms. It also works to educate society about extraterrestrial intelligence. The center publishes numerous position papers, such as "Understanding UFO Secrecy" and "Abductions: Not All That Glitters Is Gold," as well as field reports on UFOs.

Center for UFO Studies (CUFOS)
2457 W. Peterson Ave., Chicago, IL 60659
(773) 271-3611
e-mail: infocenter@cufos.org • website: www.cufos.org

CUFOS is a nonprofit scientific organization dedicated to the continuing examination and analysis of the UFO phenomenon. The center acts as a clearinghouse for the reporting and researching of UFO experiences. It publishes the quarterly *International UFO Reporters* and the *Journal of UFO Studies*.

Citizens Against UFO Secrecy, Inc. (CAUS)
PO Box 20351, Sedona, AZ 86341-0351
(602) 818-8248
website: www.caus.org

CAUS believes that extraterrestrial intelligence is in contact with Earth and that there is a campaign of secrecy to conceal this knowledge. Its goals are to educate and enlighten the public about this cover-up and to fund further research into extraterrestrial contact with Earth. It publishes the quarterly newsletter *Just Cause* and the book *Clear Intent.*

Committee for the Scientific Investigation of Claims of the Paranormal (CSICOP)
PO Box 703, Amherst, NY 14226
(716) 636-1425 • fax: (716) 636-1733
e-mail: info@csicop.org • website: www.csicop.org

Established in 1976, the committee is a nonprofit scientific and educational organization that encourages the critical investigation of paranormal and fringe-science claims from a scientific point of view. It disseminates factual information about the results of such inquiries to the scientific community and the public. CSICOP publishes *Skeptical Inquirer* magazine, the children's book *Bringing UFOs Down to Earth,* and bibliographies of other published materials that examine claims of the paranormal.

Intruders Foundation (IF)

PO Box 30233, New York, NY 10011
(212) 645-5278 • fax: (212) 352-1778
e-mail: ifcentral@aol.com • website: www.intrudersfoundation.org/inside.html

The Intruders Foundation was established by ufologist Budd Hopkins as a forum to provide sympathetic help and support to those who have experienced alien abductions. IF also investigates and researches the abduction phenomenon. Among the foundation's publications are the quarterly *IF Newsletter* and occasional special reports, including *The UFO Phenomenon and the Suicide Cults—An Ideological Study.*

Mutual UFO Network (MUFON)

103 Oldtowne Rd., Seguin, TX 78155-4099
(210) 379-9216 • fax: (210) 372-9439
e-mail: mufonq@aol.com • website: www.mufon.org
Canadian website: www.renaissoft.com/ufocanada/

MUFON is the world's largest civilian UFO research organization. MUFON documents, investigates, and studies cases of UFO sightings and alien encounters. It also holds conferences and symposiums on UFO-related issues. Its members include well-known ufologists and abductees, as well as physicians, psychiatrists, psychologists, astronomers, theologians, engineers, and other scientific professionals. It publishes the *MUFON UFO Journal* and *Symposium Proceedings,* which contains reports on its conferences.

National UFO Reporting Center

PO Box 45623, University Station, Seattle, WA 98145
UFO report hotline: (206) 722-3000
website: www.ufocenter.com

The center serves as a headquarters for reporting possible UFO sightings. Such reports are recorded and disseminated for objective research and information purposes. The center maintains an online database of all UFO reports.

SETI Institute

2035 Landings Dr., Mountain View, CA 94043
(650) 961-6633 • fax: (650) 961-7099
website: www.seti-inst.edu

The SETI Institute is a scientific organization that conducts the world's most comprehensive search for extraterrestrial intelligence. The radio telescopes used in its Project Phoenix scan nearby stars, searching for radio signals from other planets. Its goal is to map the origin, prevalence, and distribution of life in the universe. The institute publishes the newsletter *SETI News.*

Skeptics Society

PO Box 338, Altadena, CA 91001
(818) 794-3119 • fax: (818) 794-1301
e-mail: skepticmag@aol.com • website: www.skeptic.com

The society is composed of scholars, scientists, and historians who promote the use of scientific methods to scrutinize such nonscientific beliefs as religion, superstition, mysticism, and New Age tenets. It is devoted to the investigation of extraordinary claims and revolutionary ideas and to the promotion of science and critical thinking. The society publishes the quarterly *Skeptic Magazine.*

BIBLIOGRAPHY

Books

Joel Achenbach
Captured by Aliens: The Search for Life and Truth in a Very Large Universe. New York: Simon and Schuster, 1999.

Robert E. Bartholomew and George S. Howard
UFOs and Alien Contact: Two Centuries of Mystery. Amherst, NY: Prometheus, 1998.

Mark J. Carlotto
The Martian Enigma: A Closer Look. Berkeley, CA: North Atlantic, 1997.

Andrew J.H. Clark
Aliens: Can We Make Contact with Extraterrestrial Intelligence? New York: Fromm International, 1999.

Philip J. Corso with William J. Birnes
The Day After Roswell. New York: Pocket, 1997.

Jodi Dean
Aliens in America: Conspiracy Cultures from Outerspace to Cyberspace. Ithaca, NY: Cornell University Press, 1998.

Bill Fawcett, ed.
Making Contact: A Serious Handbook for Locating and Communicating with Extraterrestrials. New York: Morrow, 1997.

Kendrick Frazier, Barry Karr, and Joel Nickell, eds.
The UFO Invasion: The Roswell Incident, Alien Abductions, and Government Cover-ups. Amherst, NY: Prometheus, 1997.

Donald Goldsmith
The Hunt for Life on Mars. New York: Dutton, 1997.

Timothy Good
Alien Base: Earth's Encounters with Extraterrestrials. London: Century, 1998.

Richard F. Haines
CE-5: Close Encounters of the Fifth Kind: 242 Case Files Exposing Alien Contact. Naperville, IL: Sourcebooks, 1999.

Paul Halpern
The Quest for Alien Planets: Exploring Worlds Outside the Solar System. New York: Plenum, 1997.

Michael Hessemann and Philip Mantle
Beyond Roswell: The Alien Autopsy Film, Area 51, and the U.S. Government Cover-up of UFOs. London: Michael O'Mara, 1997.

J. Allen Hynek
The Hynek UFO Report. New York: Barnes and Noble, 1997.

David Michael Jacobs
The Threat: The Secret Alien Agenda. New York: Simon and Schuster, 1998.

Philip J. Klass
The Real Roswell Crashed-Saucer Cover-up. Amherst: Prometheus, 1997.

David Koerner and Simon LeVay
Here Be Dragons: The Scientific Quest for Extraterrestrial Life. New York: Oxford University Press, 2000.

Roger K. Leir

The Aliens and the Scalpel: Scientific Proof of Extraterrestrial Implants in Humans. Columbus, NC: Granite, 1998.

Michael D. Lemonick

Other Worlds: The Search for Life in the Universe. New York: Simon and Schuster, 1998.

John E. Mack

Passport to the Cosmos: Human Transformation and Alien Encounters. New York: Crown, 1999.

Jim Marrs

Alien Agenda: Investigating the Extraterrestrial Presence Among Us. New York: HarperCollins, 1997.

Terry Matheson

Alien Abductions: Creating a Modern Phenomenon. Amherst, NY: Prometheus, 1998.

James McAndrew

The Roswell Report: Case Closed. Washington, DC: U.S. Air Force, 1997.

Paul Meehan

Saucer Movies: A UFOlogical History of the Cinema. Lanham, MD: Scarecrow, 1998.

Barry R. Parker

Alien Life: The Search for Extraterrestrials and Beyond. New York: Plenum, 1998.

Phil Patton

Dreamland: Travels Inside the Secret World of Roswell and Area 51. New York: Villard, 1998.

Kevin D. Randle

The Randle Report: UFOs in the '90s. New York: M. Evans and Cole, 1997.

Benson Saler, Charles A. Ziegler, and Charles B. Moore

UFO Crash at Roswell: The Genesis of a Modern Myth. Washington, DC: Smithsonian Institution, 1997.

Robert Shapiro

Planetary Dreams: The Quest to Discover Life Beyond Earth. New York: Wiley, 1999.

G. Seth Shostak

Sharing the Universe: Perspectives on Extraterrestrial Life. Berkeley, CA: Berkeley Hills, 1998.

Whitley Strieber

Confirmation: The Hard Evidence of Aliens Among Us. New York: St. Martin's, 1998.

Peter D. Ward and Donald Brownlee

Rare Earth: Why Complex Life Is Uncommon in the Universe. New York: Copernicus, 1999.

David Wilkinson

Alone in the Universe? Aliens, the X-Files, and God. Downers Grove, IL: InterVarsity, 1997.

Periodicals

Robert E. Bartholomew

"Before Roswell: The Meaning Behind the Crashed-UFO Myth," *Skeptical Inquirer,* May/June 1998.

Janet Bergmark

"My Life with an Alien," *Fate,* September 1997. Available from PO Box 1940, 170 Future Way, Marion, OH 43305-1940.

Richard Boylan

"Inside Revelations on the UFO Cover-Up," *Nexus,* April/May 1998. Available from PO Box 22034, Tulsa, OK 74121.

William J. Broad

"CIA Admits Government Lied About UFO Sightings," *New York Times,* August 3, 1997.

William J. Broad   "Wanna See a Real Live Martian? Try the Mirror," *New York Times*, March 14, 1999.

Christianity Today   "To Heaven on a UFO?" May 19, 1997.

Paul Davies   "Is There Life Out There?" *Wall Street Journal*, September 24, 1999.

Preston E. Dennett   "Alien Healings: The Medical Evidence," *Nexus*, October/November 1998.

David J. Eicher   "Are We Alone?" *Astronomy*, November 1999.

Randy Fitzgerald   "UFOs: A Second Look," *Reader's Digest*, May 1999.

Jeff Greenwald   "To Infinity . . . and Beyond!" *Wired*, July 1998. Available from PO Box 55689, Boulder, CO 80322-5689.

Bruce Handy   "Roswell or Bust," *Time*, June 23, 1997.

Lawrence M. Krauss   "Stop the Flying Saucer, I Want to Get Off," *New York Times*, February 22, 1999.

Andrew J. LePage and   "SETI Searches Today," *Sky and Telescope*, December
Alan M. MacRobert   1998.

Art Levine   "A Little Less Balance, Please," *U.S. News & World Report*, July 14, 1997.

Charles McGrath   "It Just Looks Paranoid," *New York Times Magazine*, June 14, 1998.

Lisa Miller   "If We're Not Alone," *Wall Street Journal*, January 1, 2000.

Joe Nickell   "Alien Implants: The New 'Hard Evidence,' " *Skeptical Inquirer*, September/October 1998.

Govert Schilling   "The Chance of Finding Aliens: Re-evaluating the Drake Equation," *Sky and Telescope*, December 1998.

Robert Sheaffer   "The Truth Is, They Never Were 'Saucers,' " *Skeptical Inquirer*, September/October 1997.

Richard A. Shweder   "How We Down Here View What's Out There," *New York Times*, August 24, 1997.

Scott Smith   "Alien Seed: Is Our Technology Really Ours?" *Fate*, March 1999.

Dawn Stover   "The Great Big Telescope," *Popular Science*, January 2000.

Keith Thompson   "Crop Circles: Stalking a Grain of Truth," *Ions Noetic Sciences Review*, April-July 1999. Available from 475 Gate Five Rd., Suite 300, Sausalito, CA 94965.

Time   Special section on space and science, April 10, 2000.

Jim Wilson   "The Secret CIA UFO Files," *Popular Mechanics*, November 1997.

David Wise   "Big Lies and Little Green Men," *New York Times*, August 8, 1997.

INDEX